What does this painting have to do with math?

An intersection in Paris on a gray, rainy day is the subject of this atmospheric Impressionist painting. Gustave Caillebotte creates depth in this scene by using perspective and proportion in a variety of ways, including by placing large figures in the foreground and smaller ones in the distance. Imagine there is a coordinate grid on the building in the background. How might you determine the distance from the front of the building to the back by using the coordinate plane?

On the cover

Paris Street; Rainy Day, 1877
Gustave Caillebotte, French, 1848–1894
Oil on canvas
The Art Institute of Chicago, Chicago, IL, USA

Gustave Caillebotte (1848–1894). *Paris Street; Rainy Day*, 1877. Oil on canvas, 212.2 × 276.2 cm (83 1/2 × 108 3/4 in). Charles H. and Mary F. S. Worcester Collection (1964.336). The Art Institute of Chicago, Chicago, IL, USA. Photo Credit: The Art Institute of Chicago/Art Resource, NY

EUREKA MATH²

GREAT
MINDS
™

Great Minds® is the creator of *Eureka Math*®,
Wit & Wisdom®, *Alexandria Plan*™, and *PhD Science*®.

Published by Great Minds PBC.
greatminds.org

Printed in the USA
B-Print

2 3 4 5 6 7 8 9 10 11 QDG 27 26 25 24 23

ISBN 978-1-64497-119-2

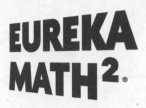

EUREKA MATH²®

A Story of Ratios®

Ratios and Rates ▸ 6

LEARN

Contents

Ratios, Rates, and Percents

Resources

Paint Ratios

The magic of ratios is that they let us compare quantities.

Quantities like "gallons of red paint" and "gallons of white paint."

Or "cups of flour" and "cups of sugar."

Or "number of adults" and "number of students."

But be careful. When using a ratio, we need to specify *which* quantities we're comparing! Otherwise, we may fall victim to disastrous errors.

A 2 to 1 ratio of gallons of red paint to gallons of white paint will yield a pleasing shade of pink. But a 2 to 1 ratio of the volume of paint cans *to the volume of the house itself* will yield an enormous pile of paint cans, twice the size of the house you're trying to paint.

That's not recommended, unless you're trying to build a new, larger house out of paint cans. In that case, go for it!

Name _____ Date _____

Jars of Jelly Beans

Explore

Focus question:

Why is The number of Jelly Beans uneven.

how many Jelly Beans do you think could fit in the Jars.

Help comeing

Cost of Jelly Beans

Focus question:

EXIT TICKET 1

Name _____ Date _____

How did your estimate for the number of jelly beans that could fit in each jar compare to the actual number of jelly beans that could fit in each jar? Explain why your estimate was different from the actual number.

300

800

_____   _____
Name   Date

1. How did you use mathematical tools to estimate the number of jelly beans in each jar?

2. You want to estimate how long it will take a hose to fill a 5-gallon bucket with water.

   a. What information do you need to answer your question?

   b. What assumptions might you need to make to answer your question?

   c. You use a stopwatch to record the time it takes the hose to fill an 8-ounce jar. You find that it takes 2.5 seconds. With this information, what is a reasonable estimate for how long it will take the hose to fill the 5-gallon bucket? Explain your reasoning.

Remember

For problems 3–5, multiply.

3. 471×3

4. 809×4

5. 975×5

6. Convert 24 yards to feet.

7. Which statements correctly describe the equation $12 \times 15 = 180$? Choose all that apply.

   A. 180 is 15 more than 12.

   B. 180 equals 15 times as many as 12.

   C. 180 equals 12 times as many as 15.

   D. 180 is 12 more than 15.

   E. 180 represents 12 groups of 15.

Name _____   Date _____

Introduction to Ratios

A New Language

1. Lisa has 9 tokens. Toby has 13 tokens. Which statements describe the relationship between the two quantities? Choose all that apply.

   A. Lisa has $\frac{9}{13}$ as many tokens as Toby.

   B. Lisa has $\frac{13}{9}$ times as many tokens as Toby.

   C. A ratio that relates the number of tokens Lisa has to the number of tokens Toby has is $9:13$.

   D. A ratio that relates the number of tokens Lisa has to the number of tokens Toby has is $13:9$.

   E. For every 9 tokens Lisa has, Toby has 13 tokens.

   F. For every 9 tokens Toby has, Lisa has 13 tokens.

From Tokens to Tea

2. A recipe for lemonade calls for 2 lemons and 5 cups of water. Which statements describe the relationship between the two quantities? Choose all that apply.

   A. A ratio that relates the number of lemons to the number of cups of water is 2 to 5.

   B. A ratio that relates the number of cups of water to the number of lemons is 5 to 2.

   C. A ratio that relates the number of cups of water to the number of lemons is 2 to 5.

   D. For every 5 cups of water, there are 2 lemons.

   E. For every 2 cups of water, there are 5 lemons.

   F. There is $2\frac{1}{2}$ times as much water as lemons.

3. To make light blue paint, Ryan mixes 2 ounces of white paint with 6 ounces of blue paint. For parts (a)–(e), fill in the blanks.

   a. A ratio that relates the number of ounces of white paint to the number of ounces of blue paint is _____

   b. A ratio that relates the number of ounces of blue paint to the number of ounces of white paint is _____

   c. For every _____ ounces of white paint, Ryan mixes 6 ounces of blue paint.

   d. For every 1 ounce of white paint, Ryan mixes _____ ounces of blue paint.

   e. Ryan uses _____ times as much blue paint as white paint.

Name _____ Date _____

Consider the cans of blue paint and the cans of red paint.

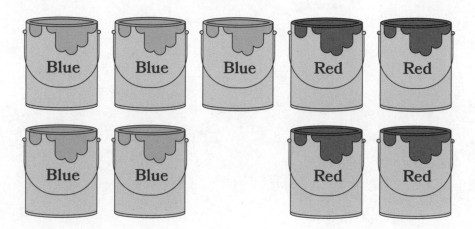

For parts (a)–(d), fill in the blank.

a. A ratio that relates the number of cans of blue paint to the number of cans of red paint is _____.

b. A ratio that relates the number of cans of red paint to the number of cans of blue paint is _____.

c. There are _____ times as many cans of blue paint as cans of red paint.

d. For every _____ cans of blue paint, there are _____ cans of red paint.

RECAP 2

Name _____ Date _____

Introduction to Ratios

In this lesson, we

- wrote ratios to relate two quantities.

- used multiplicative comparison language to compare two quantities.

- used ratio language to compare two quantities.

> **Terminology**
>
> A **ratio** is an ordered pair of numbers that are not both zero.
>
> A ratio can be written as A to B or $A : B$.

Examples

1. Consider the collection of shapes shown. Which statements correctly describe the collection of shapes? Choose all that apply.

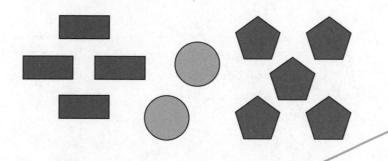

A. A ratio that relates the number of orange rectangles to the number of blue pentagons is $5 : 4$.

> The order in which the quantities are described tells us the order of the numbers in the ratio.
>
> So the ratio of the number of orange rectangles to the number of blue pentagons is $4 : 5$, not $5 : 4$.

B. There are $2\frac{1}{2}$ times as many pentagons as circles.

> A ratio that relates the number of pentagons to the number of circles is $5 : 2$. That means there are $\frac{5}{2}$, or $2\frac{1}{2}$, times as many pentagons as circles.

C. For every 1 circle, there are 2 rectangles.

> For every 2 circles, there are 4 rectangles. That means there are twice as many rectangles as circles.
>
> So for every 1 circle, there are 2 rectangles.

D. There are $\frac{1}{2}$ as many orange rectangles as yellow circles.

> For every 4 orange shapes, there are 2 yellow shapes. That means there are 2 times as many orange rectangles as yellow circles, not $\frac{1}{2}$ as many.

2. There are 12 students who take orchestra class. There are 4 times as many students who take band class as students who take orchestra class.

   a. Write a ratio that relates the number of students who take orchestra class to the number of students who take band class.

   A ratio that relates the number of students who take orchestra class to the number of students who take band class is $12 : 48$.

   > The number of students who take band class is 48 because $4 \times 12 = 48$.

   b. Scott uses ratio language to describe the ratio from part (a). He says that for every 1 student who takes orchestra class, there are 4 students who take band class. Is Scott correct? Why?

   Yes. Scott is correct because there are 4 times as many students who take band class as students who take orchestra class and $1 \times 4 = 4$.

   > There are $\frac{1}{4}$ as many students who take orchestra class as students who take band class.

Name _____ Date _____

1. A smoothie recipe calls for bananas and strawberries in the ratio represented by the picture.

For parts (a)–(d), fill in the blanks.

A. A ratio that relates the number of strawberries to the number of bananas is _____.

B. There are _____ times as many strawberries as bananas.

C. For every _____ bananas, there are 7 strawberries.

D. There are _____ times as many bananas as strawberries.

2. Consider the collection of shapes shown. Which statements correctly describe the collection of shapes? Choose all that apply.

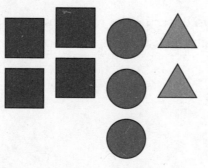

A. A ratio that relates the number of red squares to the number of blue circles is 4 : 3.

B. There are 2 times as many triangles as squares.

C. There are $\frac{1}{2}$ as many triangles as squares.

D. For every 1 triangle, there are 2 squares.

E. A ratio that relates the number of squares to the number of circles is 3 : 4.

3. Sasha says there are $1\frac{1}{2}$ times as many circles as triangles in the picture in problem 2. Is she correct? Explain.

4. At an animal shelter, 9 dogs and 15 cats are ready for adoption. Fill in the blanks to make the statements true.

    a. For every _____ dogs, there are 15 cats.

    b. For every 3 dogs, there are _____ cats.

    c. There are _____ times as many cats as dogs.

5. Students at a middle school take an elective during the last hour of the school day. There are 11 students who take an art class. There are 3 times as many students who take a music class as students who take an art class.

    a. What is a ratio that relates the number of students who take a music class to the number of students who take an art class?

    b. Kayla uses ratio language to describe the ratio from part (a). She says that for every 3 students who take a music class, there is 1 student who takes an art class. Is Kayla correct? Why?

Remember

For problems 6–8, multiply.

6. $1,312 \times 3$

7. $2,214 \times 4$

8. $5,631 \times 5$

9. Convert 5 hours to minutes.

10. Each model is divided into equal sections. Which models have a shaded portion that represents the fraction $\frac{1}{2}$? Choose all that apply.

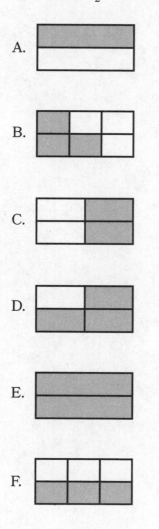

A.

B.

C.

D.

E.

F.

_____   _____
Name   Date

Ratios and Tape Diagrams

Shirts and Seating Arrangements

1. In the flower case at the supermarket, there are 5 bouquets of red roses and 4 bouquets of pink roses. Each bouquet is half a dozen roses.

    a. Write and explain the meaning of two ratios that could represent this situation.

    b. A florist uses all the roses to make new bouquets. Each bouquet is an identical mixture of red roses and pink roses. How many bouquets can he make? How many red roses and how many pink roses are in each bouquet?

    c. Use ratio language to describe the relationship between the number of red roses and the number of pink roses in each bouquet in part (b).

Using Tape Diagrams to Represent Ratios

2. The company also offers printed photos that have a width of 5 inches and a height of 7 inches.

5 in Width

Height 7 in

a. Draw a tape diagram to represent the ratio of the width of the photo to the height of the photo.

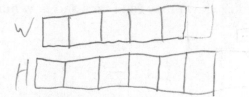

b. What does 1 unit in the tape diagram represent?

c. What is the ratio of the width of the photo to the height of the photo?

d. What is the ratio of the height of the photo to the width of the photo?

e. The height of the photo is $\frac{7}{5}$ times as much as the width of the photo.

Fraction

f. What is the ratio of the width of the photo to the perimeter of the photo?

3. Many older televisions have the ratio of width to height that is represented in the tape diagram.

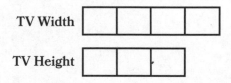

a. What is the ratio of the width of the television to the height of the television?

b. According to the tape diagram, what could be a possible width and height of an older television?

Name _____ Date _____

Tyler has 24 quarters and 6 dimes.

a. Write and explain the meaning of two ratios that could represent this situation.

b. Tyler puts all the coins in bags. Every bag has the same number of quarters and the same number of dimes. How many bags of coins can Tyler make? How many quarters and how many dimes are in each bag?

c. Use ratio language to describe the relationship between the number of quarters and the number of dimes in each bag from part (b).

d. Use ratio language to describe the relationship between the number of boxes of raisins and the number of bags of popcorn.

Sample: For every 2 boxes of raisins, there are 3 bags of popcorn.

e. Draw a tape diagram that represents the ratio of the number of boxes of raisins to the number of bags of popcorn in one gift basket.

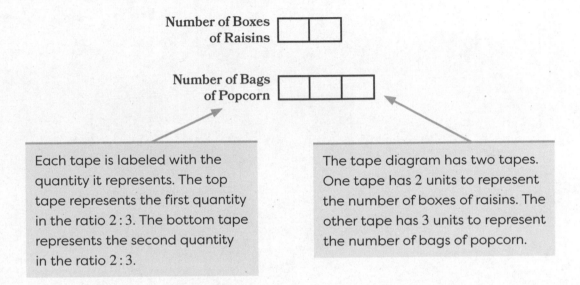

Each tape is labeled with the quantity it represents. The top tape represents the first quantity in the ratio 2 : 3. The bottom tape represents the second quantity in the ratio 2 : 3.

The tape diagram has two tapes. One tape has 2 units to represent the number of boxes of raisins. The other tape has 3 units to represent the number of bags of popcorn.

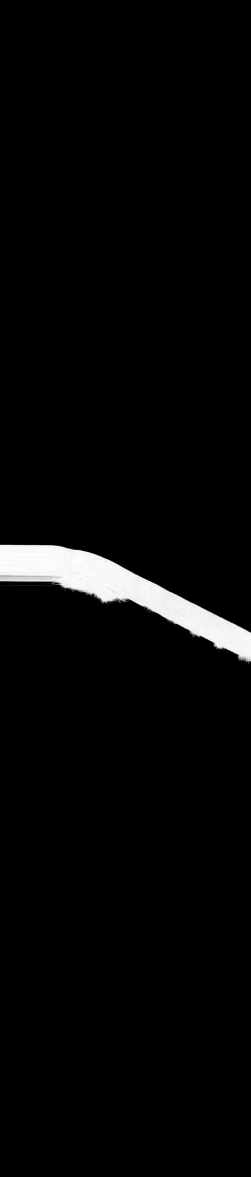

Name _____ Date _____

1. The local animal shelter has 8 bags of dog toys and 10 bags of cat toys.

   a. Write a ratio that relates the number of bags of dog toys to the number of bags of cat toys.

   b. What could the ratio $10 : 18$ represent?

   c. There are 6 toys in each bag of pet toys. What is the ratio of the total number of dog toys to the total number of cat toys?

   d. Explain how the ratios you wrote in part (a) and part (c) represent the same situation.

2. Sana prepares party favor bags. She has 16 stickers and 24 crayons to put in the bags.

   a. Write a ratio that relates the number of stickers to the number of crayons.

   b. Sana prepares 8 party favor bags. The ratio of the number of stickers to the number of crayons in each bag is the same. How many stickers are in each bag? How many crayons are in each bag?

c. Use ratio language to describe the relationship between the number of stickers and the number of crayons in each party favor bag.

d. Draw a tape diagram to represent the ratio of the number of stickers to the number of crayons in each party favor bag.

3. A pet store has 10 cockatiels and 15 parakeets.

a. Write a ratio that relates the number of cockatiels to the number of parakeets.

b. The pet store owner wants to group the birds so that cockatiels and parakeets are not placed together in a cage. He wants each cage to have the same number of birds. How can the pet store owner group the birds?

c. Use the groups from part (b) to write the ratio of the number of groups of cockatiels to the number of groups of parakeets.

d. Use ratio language to describe the relationship between the number of cockatiels and the number of parakeets.

e. Explain how the ratios you wrote in part (a) and part (c) represent the same situation.

4. Of the students who help clean a park, 12 are basketball players and 18 are band members. Which statements are correct? Choose all that apply.

   A. There are $1\frac{1}{2}$ times as many band members as basketball players.

   B. The ratio of the number of band members to the number of basketball players is $12:18$.

   C. The ratio of the number of basketball players to the total number of students is $12:30$.

   D. Six groups of students could each have 2 basketball players and 3 band members.

   E. Two groups of basketball players and 3 groups of band members could each have 6 people.

   F. There are 2 basketball players for every 3 band members.

5. The tape diagram represents the ratio of the number of carnations to the number of daisies in a flower bouquet.

Number of Carnations			

   Number of Carnations

   Number of Daisies

   Use ratio language to describe the ratio that is represented in the tape diagram.

Remember

For problems 6–8, multiply.

6. $3,132 \times 3$

7. $2,416 \times 4$

8. $7,921 \times 5$

9. Use the given rule and starting number to complete each numerical pattern.

   a. Add $\frac{1}{2}$, starting with 0.

   0, _____ , _____ , _____ , _____ , _____

   b. Add 1, starting with 0.

   0, _____ , _____ , _____ , _____ , _____

10. Lacy rides her bicycle 15 miles on Sunday. This distance is 5 times as far as she rode on Saturday. Which number sentence shows how to find the number of miles Lacy rode her bicycle on Saturday?

   A. $15 + 5 = 20$

   B. $15 - 5 = 10$

   C. $15 \times 5 = 75$

   D. $15 \div 5 = 3$

LESSON 4

_____
Name

_____
Date

Exploring Ratios by Making Batches

Tiling

Batches of Paint

Name _____   Date _____

1. Tyler uses 6 triangles and 1 hexagon to create the sun shown.

a. Draw a tape diagram to represent the ratio of the number of triangles to the number of hexagons in Tyler's sun.

b. Write a ratio that relates the number of triangles to the number of hexagons in Tyler's sun.

c. Write a ratio that relates the number of hexagons to the total number of shapes in Tyler's sun.

d. Draw a tape diagram to represent the total numbers of triangles and hexagons that Tyler needs to create 2 suns.

e. Draw a tape diagram to represent the total numbers of triangles and hexagons that Tyler needs to create 3 suns.

2. Riley's recipe for salad dressing calls for 2 tablespoons of vinegar for every 3 tablespoons of olive oil.

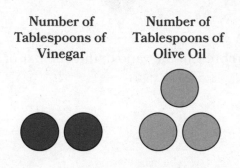

Number of Tablespoons of Vinegar   **Number of Tablespoons of Olive Oil**

a. Draw a tape diagram to represent the ratio of the number of tablespoons of vinegar to the number of tablespoons of olive oil.

b. If Riley uses 4 tablespoons of vinegar, how many tablespoons of olive oil does she need? Draw a tape diagram to show your thinking.

c. If Riley uses 9 tablespoons of olive oil, how many tablespoons of vinegar does she need? Draw a tape diagram to show your thinking.

3. A company makes gift baskets that each include 2 bottles of maple syrup and 5 chocolate bars.

a. How many bottles of maple syrup and how many chocolate bars does the company need to make 6 baskets?

b. If the company uses 20 chocolate bars in gift baskets, how many bottles of maple syrup does it use?

4. Sasha mixes 4 tablespoons of white paint with 5 tablespoons of red paint to create pink paint.

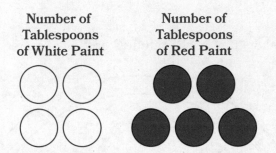

| Number of Tablespoons of White Paint | Number of Tablespoons of Red Paint |

Which mixtures create the same shade of pink paint? Choose all that apply.

A. 4 cups of white paint and 5 cups of red paint

B. 3 cups of white paint and 4 cups of red paint

C. 6 tablespoons of white paint and 7 tablespoons of red paint

D. 8 tablespoons of white paint and 10 tablespoons of red paint

E. 2 tablespoons of white paint and $2\frac{1}{2}$ tablespoons of red paint

Remember

For problems 5–7, multiply.

5. 24×14

6. 33×25

7. 42×36

8. Write two ratios that relate the number of blue ovals and the number of red triangles. Explain the meaning of each ratio.

9. Choose the true statement.

   A. All rectangles are squares because all rectangles have four equal sides.

   B. All rhombuses are squares because all rhombuses have four right angles.

   C. All parallelograms are quadrilaterals because all parallelograms have four sides.

   D. All trapezoids are parallelograms because all trapezoids have two pairs of parallel sides.

Emanuel

Name

Date

Equivalent Ratios

1. A bouquet of roses and daisies has a total of 105 flowers. For every 2 roses, there are 3 daisies.

How many roses are in the bouquet? How many daisies are in the bouquet?

55 : 58

Identifying and Writing Equivalent Ratios

2. Write two ratios: one that is equivalent to $3:5$ and one that is not equivalent to $3:5$. Justify your reasoning.

Using Equivalent Ratios to Solve Problems

For problems 3 and 4, the bouquets have 2 roses for every 3 daisies. Label and use the tape diagrams to answer the questions.

3. If a bouquet has 16 roses, how many daisies are in the bouquet? What is the total number of flowers in the bouquet?

Number of Roses | 8 | 8 |

Number of Daisies | | | |

Handwritten: 16 roses, 8×5, 8×2, 40, 16+2a, 24 Daisies

4. If a bouquet has a total of 105 flowers, how many roses are in the bouquet? How many daisies are in the bouquet?

Number of Roses | | |

Number of Daisies | | | |

5. A board game includes tiles. Each tile is labeled with a letter. There are 3 tiles labeled A for every 4 tiles labeled E. If there are 9 tiles labeled A, how many tiles are labeled E?

6. A recipe to make bubbles consists of water and dishwashing soap. A ratio that relates the number of cups of water to the number of cups of dishwashing soap is 6 : 1. If a bubble mixture has 4 cups of dishwashing soap, how many cups of water does it have?

7. Sasha and Julie sell water bottles to raise money for their school. A ratio that relates the number of water bottles Sasha sells to the number of water bottles Julie sells is 5 : 2. Together, they sell 63 water bottles. How many water bottles does Julie sell?

8. The mixture in a container of hummingbird food consists of sugar and water. There are 2 ounces of sugar for every 8 ounces of water. How many ounces of sugar are in a 50-ounce container of hummingbird food?

9. An amusement park has kiddie rides and thrill rides. For every 2 kiddie rides, there is 1 thrill ride. If the amusement park has 11 thrill rides, what is the total number of rides at the amusement park?

10. A test has multiple-choice questions and essay questions. A ratio that relates the number of multiple-choice questions on the test to the number of essay questions on the test is 4 : 1. The test has 32 multiple-choice questions. What is the total number of questions on the test?

Name

Date

1. Show that the ratio 5 : 6 is equivalent to the ratio 35 : 42.

2. There are 3 red markers for every 4 blue markers in an art set. The total number of red markers and blue markers in the art set is 84.

    a. Represent this situation by using a tape diagram.

    b. How many red markers are in the art set?

    c. How many blue markers are in the art set?

RECAP **5**

Name   Date

Equivalent Ratios

In this lesson, we

- identified and wrote equivalent ratios.

- represented equivalent ratios with tape diagrams.

- determined unknown quantities by using equivalent ratios.

> **Terminology**
>
> Two ratios $A : B$ and $C : D$ are **equivalent ratios** if there is a nonzero number c such that $C = c \times A$ and $D = c \times B$.

Examples

1. Show that the ratio $5 : 2$ is equivalent to the ratio $20 : 8$.

The ratio $5 : 2$ is equivalent to the ratio $20 : 8$ because we can multiply 4 by 5 to get 20 and multiply 4 by 2 to get 8.

2. A recipe for craft clay calls for 4 ounces of cornstarch for every 5 ounces of water.

   a. Draw a tape diagram that represents the ratio of the number of ounces of cornstarch to the number of ounces of water.

   Number of Ounces
   of Cornstarch   ☐☐☐☐

   Number of Ounces
   of Water   ☐☐☐☐☐

   > The tape diagram has 4 units that represent the number of ounces of cornstarch and 5 units that represent the number of ounces of water.

b. A mixture of craft clay has 28 ounces of cornstarch. How many ounces of water does the mixture have? Use your tape diagram from part (a) to support your answer.

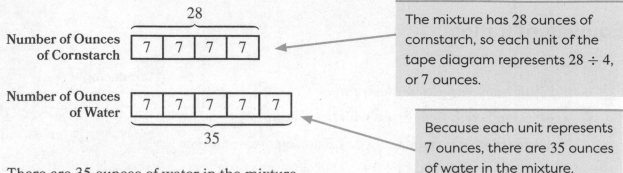

The mixture has 28 ounces of cornstarch, so each unit of the tape diagram represents 28 ÷ 4, or 7 ounces.

Because each unit represents 7 ounces, there are 35 ounces of water in the mixture.

There are 35 ounces of water in the mixture.

c. To make a total of 45 ounces of craft clay, how many ounces of cornstarch and how many ounces of water must be mixed?

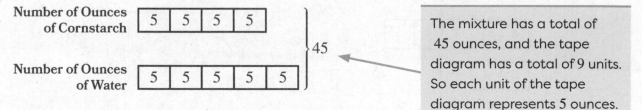

The mixture has a total of 45 ounces, and the tape diagram has a total of 9 units. So each unit of the tape diagram represents 5 ounces.

To make a total of 45 ounces of craft clay, 20 ounces of cornstarch and 25 ounces of water must be mixed.

d. A mixture of craft clay has 15 ounces of water. What is the total number of ounces of craft clay in this mixture?

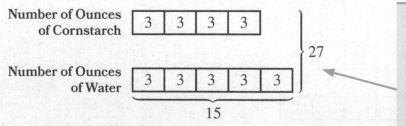

The mixture has 15 ounces of water, so each unit of the tape diagram represents 15 ÷ 5, or 3 ounces. Because the tape diagram has a total of 9 units, there is a total of 27 ounces of craft clay in this mixture.

There is a total of 27 ounces of craft clay in this mixture.

PRACTICE 5

Name _____

Date _____

1. In a rice recipe, a ratio that relates the number of cups of water to the number of cups of rice is $2 : 1$.

   a. Draw a tape diagram to represent this ratio.

   b. How many cups of water should be mixed with 4 cups of rice?

2. The tape diagram shows that the ratio $4 : 3$ is equivalent to the ratio $8 : 6$.

   | 2 | 2 | 2 | 2 |

   | 2 | 2 | 2 |

   a. Draw a tape diagram to show that the ratio $4 : 3$ is equivalent to the ratio $20 : 15$.

   b. Yuna thinks that the ratio $4 : 3$ is equivalent to the ratio $10 : 9$ because $6 + 4 = 10$ and $6 + 3 = 9$. What is Yuna's mistake? What can she do to find an equivalent ratio?

3. Consider the ratios $3:5$ and $21:D$. What must c and D be for the ratios to be equivalent?

4. The tape diagram represents the ratio of the number of free throws made to the number of free throws missed at a basketball game.

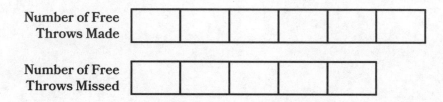

a. If this pattern continues and 12 free throws are made, what is the total number of free throw attempts in the game?

b. If this pattern continues and there are a total of 33 free throw attempts in the game, how many free throws are made? How many free throws are missed?

5. A sidewalk chalk recipe calls for 1 ounce of cornstarch for every 3 ounces of baking soda.

a. Draw a tape diagram that represents the ratio of the number of ounces of cornstarch to the number of ounces of baking soda.

b. How many ounces of cornstarch and how many ounces of baking soda are in a total of 32 ounces of sidewalk chalk?

c. A mixture of sidewalk chalk has 15 ounces of baking soda. What is the total number of ounces of sidewalk chalk in the mixture?

6. For every 4 laps that Leo runs, Tyler walks 2 laps. How many laps does Leo run if Tyler walks 10 laps?

7. A ratio that relates the number of ounces of cream cheese to the number of ounces of yogurt in a fruit dip recipe is $1 : 1$. How many ounces of cream cheese and how many ounces of yogurt are in 16 ounces of fruit dip?

8. Adults and children attend a circus. For every 2 adults at the circus, there are 3 children. If 275 people attend the circus, how many of them are children?

9. Draw a line from the left column to the right column to connect representations of equivalent ratios.

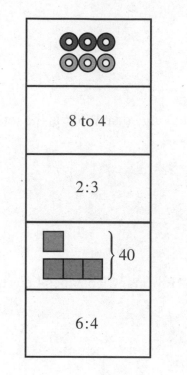
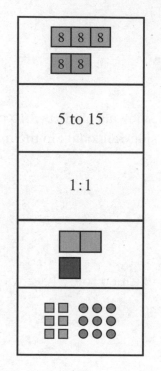

Remember

For problems 10–12, multiply.

10. 37×24　　　　　11. 52×46　　　　　12. 73×32

13. There are 20 baseball cards and 10 football cards on display at a sports museum. Noah says that the ratio of the number of baseball cards to the number of football cards is 2 : 1. Sana says that the ratio of the number of baseball cards to the number of football cards is 4 : 2. How can they both be right?

14. Toby uses 3 cups of milk to make 4 batches of pancakes. How many cups of milk does he need to make 1 batch of pancakes?

Liking "Like"

Does it seem silly to apply mathematics to language? It's not! By counting words, people have been able to solve all kinds of mysteries, such as taking anonymous documents and figuring out who wrote them. One of the methods of doing this is to count which words the author uses most often.

What word do you use the most often? Is it *the*, or perhaps *I*, or maybe a filler word such as *um* or *like*? What do you think is the ratio between the numbers of times you use this word and other words?

Try counting and see what you find!

Name _____ Date _____

Ratio Tables and Double Number Lines

Organizing Equivalent Ratios

1. Consider the nutrition facts for 1 packet of sugar. Complete the table.

Nutrition Facts

Serving Size	1 Packet
Amount Per Serving	
Calories	**15**

	%Daily Value*
Total Fat 0g	0%
Saturated Fat 0g	0%
Trans Fat 0g	
Cholesterol 0mg	0%
Sodium 0mg	0%
Total Carbohydrate 4g	1%
Dietary Fiber 0g	0%
Sugars 4g	
Protein 0g	
Vitamin A	0%
Vitamin B	0%
Vitamin C	0%
Vitamin D	0%

*Percent Daily Values are based on a 2,000 calorie diet.

Number of Packets of Sugar	Number of Grams of Sugar
1	4
2	8
3	12
4	16
5	20

Using Ratio Tables and Double Number Lines to Solve Problems

2. For every 12 ounces of soda, there are 40 grams of sugar.

   a. Complete the ratio table.

Number of Ounces of Soda	Number of Grams of Sugar
12	40
24	80
36	120
48	160

   b. Use the completed ratio table from part (a) to create a double number line.

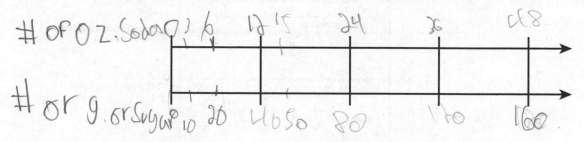

   c. How many grams of sugar are in three 12-ounce cans of this soda? How do you know?

There are 36 grams of sugar.

   d. How many grams of sugar are in half of a 12-ounce can of this soda? How do you know?

3. The graphic shows some of the different sizes of cups used to serve soda at fast food restaurants from 1955 to today. Some cups are labeled with the number of ounces of soda the cups can hold. Other cups are labeled with the number of grams of sugar in the soda the cups can hold.

Use the double number line from problem 2 to complete the ratio table.

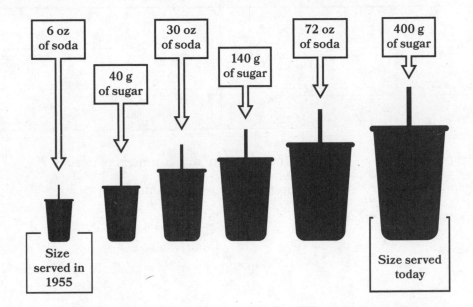

Number of Ounces of Soda	Number of Grams of Sugar
6	
	40
30	
	140
72	
	400

4. The ratio table shows the relationship between the number of cups of pretzels and the number of ounces of cereal in a snack mix recipe.

Number of Cups of Pretzels	Number of Ounces of Cereal
4	6
6	9
8	12

a. Jada says that for every 3 cups of pretzels, there are 2 ounces of cereal.

Lacy says that for every 2 cups of pretzels, there are 3 ounces of cereal.

Who is correct? Explain.

b. Complete the double number line to show the relationship between the number of cups of pretzels and the number of ounces of cereal.

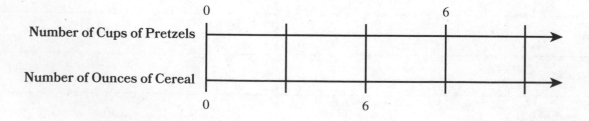

5. Leo buys fabric at a craft store. Every 2 yards of fabric costs $7.00.

   a. Create a double number line to show the relationship between possible amounts of fabric in yards and the total cost in dollars.

   b. If the total cost of the fabric is $21.00, how many yards of fabric does Leo buy?

   c. If Leo buys 1 yard of fabric, what is the total cost of the fabric?

Name _____   Date _____

1. Kelly makes bracelets by using green beads and blue beads. The tape diagram represents the ratio of the number of green beads to the number of blue beads.

Number of Green Beads [| |]

Number of Blue Beads [| | | |]

Use the tape diagram to complete the ratio table.

Number of Green Beads	Number of Blue Beads
3	
6	
9	
12	
15	

2. A recipe calls for 8 cups of water for every 16 ounces of macaroni. Complete the double number line.

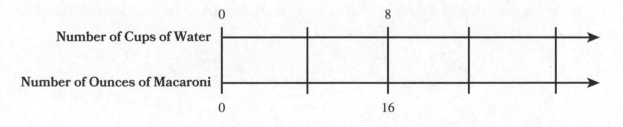

3. Blake practices piano 3 times as long as Tara does.

   a. Complete the ratio table to show possible numbers of minutes that Blake and Tara each practice piano.

Number of Minutes Blake Practices Piano	Number of Minutes Tara Practices Piano
15	
30	
	15
	20

   b. If Blake practices piano for 18 minutes, for how many minutes does Tara practice?

   c. If Tara practices piano for 30 minutes, for how many minutes does Blake practice?

4. Ryan runs 2 laps around the track in 4 minutes.

   a. Complete the double number line.

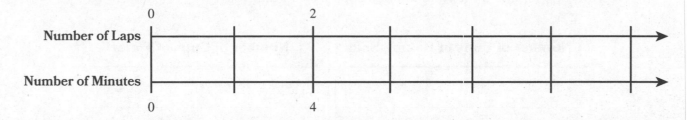

   b. If Ryan continues to run at the same pace, how many laps does he run in 10 minutes?

   c. If Ryan continues to run at the same pace, how many minutes does it take him to run 7 laps?

5. A recipe for a homemade modeling clay calls for 4 cups of baking soda for every 3 cups of water.

   a. According to this recipe, how many cups of baking soda must be mixed with 15 cups of water? Complete the ratio table to support your answer.

Number of Cups of Baking Soda	Number of Cups of Water

   b. According to this recipe, how many cups of water must be mixed with 6 cups of baking soda? Create a double number line to support your answer.

6. The Snow Ratio reports the amount of water in a given amount of snow. The ratio table shows the relationship between the number of inches of water and the number of inches of snow.

Number of Inches of Water	Number of Inches of Snow
1	10
2	20
3	30

- Tyler says that for every 10 inches of water, there is 1 inch of snow.
- Yuna says that for every 10 inches of snow, there is 1 inch of water.

Who is correct? Explain.

7. The double number line represents the ratio relationship between the number of tablespoons of vinegar and the number of drops of dishwashing soap in a homemade bug spray. Which statements are true? Choose all that apply.

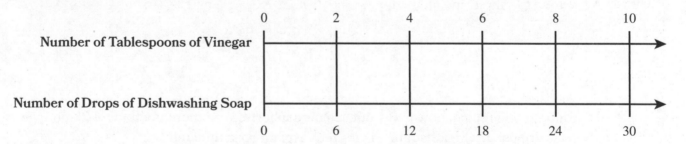

A. For every 2 tablespoons of vinegar, there are 6 drops of dishwashing soap.

B. For every 6 tablespoons of vinegar, there are 2 drops of dishwashing soap.

C. The ratio of the number of tablespoons of vinegar to the number of drops of dishwashing soap is $3:1$.

D. The ratio of the number of tablespoons of vinegar to the number of drops of dishwashing soap is $1:3$.

E. There are 5 tablespoons of vinegar for every 15 drops of dishwashing soap.

Remember

For problems 8–10, multiply.

8. $1,531 \times 20$

9. $2,347 \times 30$

10. $5,162 \times 40$

11. The ratio of the number of ounces of blueberries to the number of ounces of raspberries in a recipe is $4:3$.

a. According to this recipe, how many ounces of raspberries are there if 8 ounces of blueberries are used? Create a tape diagram to explain your thinking.

b. According to this recipe, how many ounces of raspberries are there if a total of 28 ounces of berries are used? Create a tape diagram to explain your thinking.

For problems 12–14, the coordinate plane shows the locations of Riley's and Leo's houses. It also shows the route that Riley takes to Leo's house.

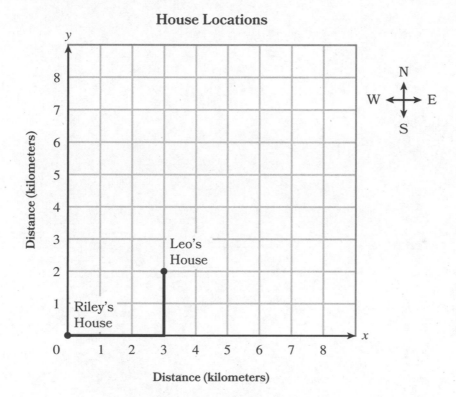

House Locations

12. Write the ordered pair that represents the location of Riley's house and the location of Leo's house.

    Riley's house: (_____,_____)

    Leo's house: (_____,_____)

13. What is the total distance of the route in kilometers that Riley takes to Leo's house?

14. Adesh's house is located 4 km east and 2 km north of Leo's house. Mark the location of Adesh's house on the given coordinate plane.

Name   Date

Graphs of Ratio Relationships

Choosing a Pet

Comparing Costs

Decision Time

Name

Date

A gorilla eats a diet of vegetables and fruits. The gorilla eats 4 vegetables for every 1 piece of fruit.

a. Complete the table.

Number of Vegetables	Number of Pieces of Fruit	Ordered Pair
4	1	(4, 1)
8		
12		
16		
20		

b. Use the ordered pairs from the table in part (a) to plot points in the coordinate plane.

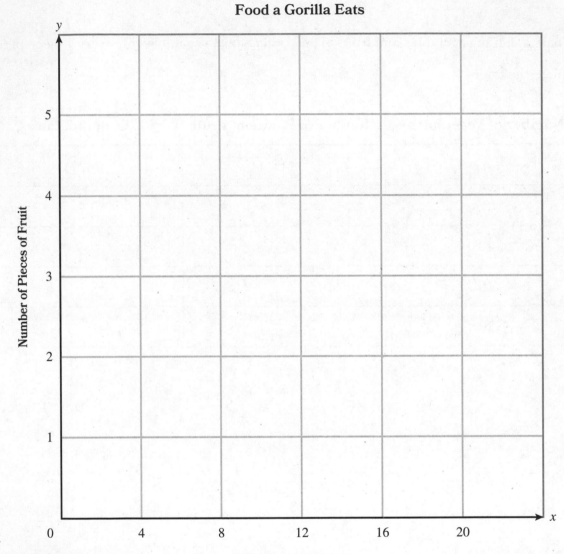

Food a Gorilla Eats

Number of Pieces of Fruit (y-axis)

Number of Vegetables (x-axis)

Name _____ Date _____

1. On a television channel, there is 1 minute of commercials for every 5 minutes of television shows.

   a. Represent the ratio with a tape diagram.

   b. Use the tape diagram from part (a) to complete the table.

Number of Minutes of Commercials	Number of Minutes of Television Shows	Ordered Pair
1	5	(1, 5)
2		
3		
4		
5		

c. Use the ordered pairs from part (b) to plot the points in the coordinate plane.

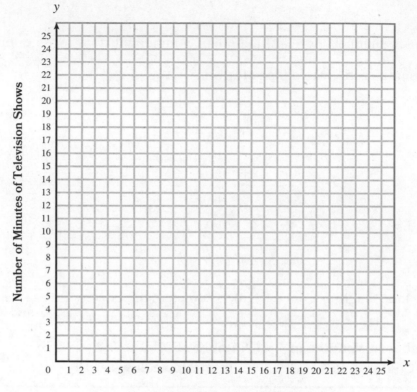

Number of Minutes of Commercials

2. The graph shows the ratio relationship between the number of tablespoons of white vinegar and the number of drops of essential oil in a recipe for a homemade cleaner.

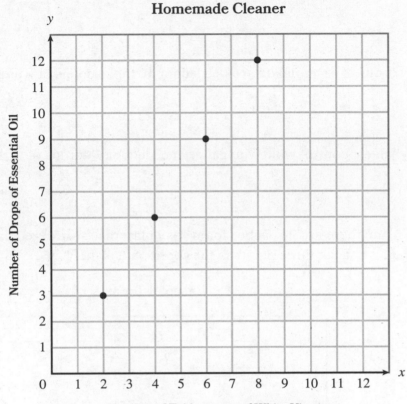

Homemade Cleaner

a. Use the graph to complete the ratio table.

Number of Tablespoons of White Vinegar	Number of Drops of Essential Oil
	3
4	
6	
	12
	15

b. What is the ratio of the number of tablespoons of white vinegar to the number of drops of essential oil?

c. How many drops of essential oil are needed for 16 tablespoons of white vinegar?

d. How many tablespoons of white vinegar are needed for 18 drops of essential oil?

3. On a baseball team, there are 2 coaches for every 9 players. Use ordered pairs from this ratio relationship to plot at least three points in the coordinate plane.

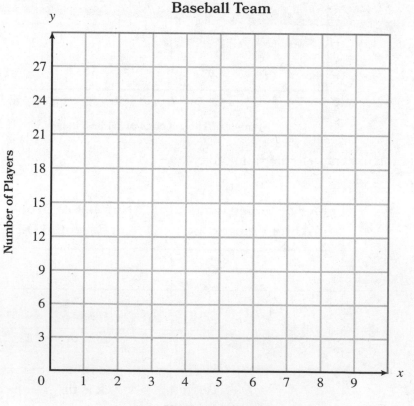

Baseball Team

4. The graph shows the ratio relationship between the number of fluid ounces and the number of tablespoons. Choose all the statements that appear to be true.

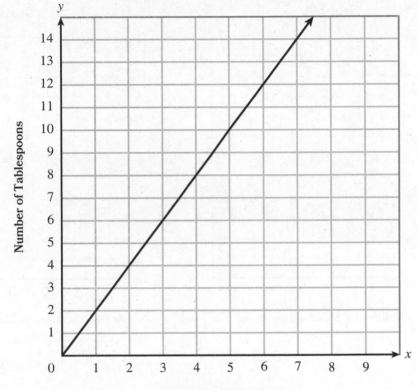

Number of Fluid Ounces

A. There are 6 fluid ounces for every 12 tablespoons.

B. There is 1 tablespoon for every 2 fluid ounces.

C. There are 5 tablespoons for every $2\frac{1}{2}$ fluid ounces.

D. There are $5\frac{1}{2}$ fluid ounces for every 11 tablespoons.

E. There are 10 tablespoons for every 20 fluid ounces.

5. Leo makes grilled cheese sandwiches for his friends. For every 2 slices of bread, he needs 1 slice of cheese. Which graph represents this ratio relationship? Explain.

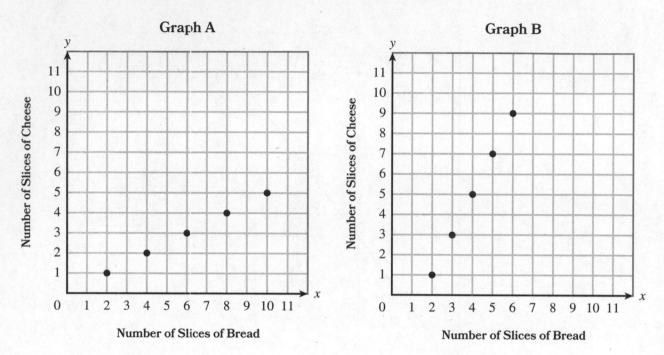

Remember

For problems 6–8, multiply.

6. $3,443 \times 20$

7. $5,165 \times 50$

8. $4,725 \times 70$

9. Julie uses green leaves and yellow leaves to make an autumn project.

   a. Complete the ratio table.

Number of Green Leaves	Number of Yellow Leaves
10	6
15	9
20	
25	

   b. What is the ratio of the number of green leaves Julie uses to the number of yellow leaves she uses?

10. Jada and Tyler attend the same middle school. Jada lives $\frac{4}{5}$ miles from the school. Tyler lives $\frac{3}{4}$ miles from the school. Choose the number sentence that correctly compares the two distances.

    A. $\frac{4}{5}$ miles $< \frac{3}{4}$ miles

    B. $\frac{4}{5}$ miles $> \frac{3}{4}$ miles

    C. $\frac{3}{4}$ miles $= \frac{4}{5}$ miles

    D. $\frac{3}{4}$ miles $> \frac{4}{5}$ miles

Name   Date

Addition Patterns in Ratio Relationships

Addition Patterns in Ratio Tables and Graphs

1. The graph represents the ratio relationship between the number of cups of orange juice and the number of cups of pineapple juice in batches of a citrus punch. Use the graph to complete the ratio table.

Batches of Citrus Punch

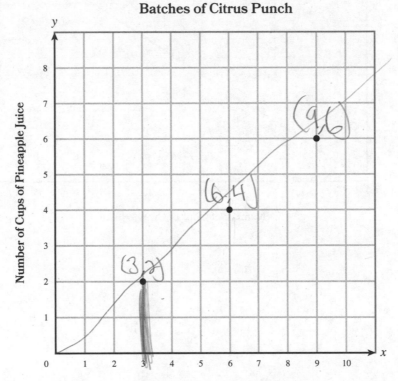

Number of Cups of Orange Juice

Number of Cups of Orange Juice	Number of Cups of Pineapple Juice
3	2
6	4
9	6

2. Use the addition patterns in the graph to identify three more ordered pairs that lie on a line with the given points. Explain what the ordered pairs represent in the ratio relationship.

Batches of Citrus Punch

Number of Cups of Orange Juice

Using Addition Patterns to Solve Problems

3. Whole milk is used to produce butter. It takes about 22 cups of whole milk to produce 1 pound of butter.

   a. Complete the ratio table.

Number of Cups of Whole Milk	Number of Pounds of Butter
22	1
44	2
66	3
88	4
110	5

   b. What is a ratio of the number of cups of whole milk used to the number of pounds of butter produced?

   44 : 2   66 : 3

   110 : 5   88 : 4

   c. Use the addition patterns in the table to fill in the blanks to describe the ratio relationship.

   For every __22__ more cups of whole milk used, there is/are __1__ more pound(s) of butter produced.

4. The graph shows the ratio relationship between the number of pounds of grapes and the total cost of the grapes at a store.

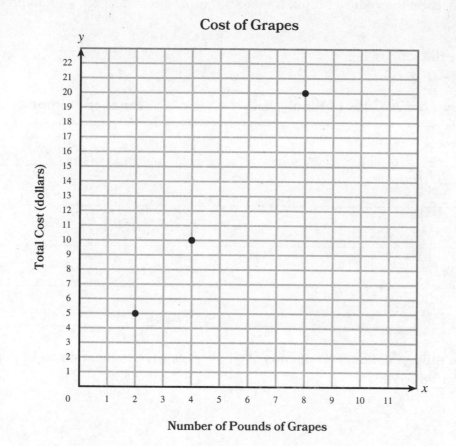

a. Describe the ratio relationship by using the addition patterns in the graph.

b. What is the total cost of 6 pounds of grapes? Use the graph to explain your answer.

c. How many pounds of grapes can someone buy with exactly $25? Use the graph to explain your answer.

5. A scientist studies a group of left-handed people. In the study, there are 10 left-handed males for every 8 left-handed females.

   a. Complete the ratio table.

Number of Left-Handed Males	10		20	25	
Number of Left-Handed Females	8	12			24

   b. Describe the ratio relationship by using the addition patterns in the table.

   c. If there are 40 left-handed males in the study, how many left-handed females are there?

Name _____   Date _____

Kayla uses green ribbon and yellow ribbon to tie bows on gifts. Use the ratio table shown to answer parts (a)–(c).

Number of Inches of Green Ribbon	Number of Inches of Yellow Ribbon
8	10
16	20
24	
32	

a. Complete the ratio table.

b. What is a ratio of the number of inches of green ribbon Kayla uses to the number of inches of yellow ribbon she uses?

c. Fill in the blanks to make a true statement.

For every _____ more inches of green ribbon Kayla uses, she uses _____ more inches of yellow ribbon.

Name   Date

1. A bag of frozen pastries includes microwave instructions. There is a ratio relationship between the number of frozen pastries and the number of seconds it takes to microwave them.

Number of Frozen Pastries	Number of Seconds
6	60
9	90

    a. Fill in the blank to make a true statement.

        For every 6 frozen pastries, it takes _____ seconds to microwave them.

    b. Fill in the blanks to make a true statement.

        For every _____ more frozen pastries, it takes _____ more seconds to microwave them.

    c. How many seconds does it take to microwave 12 frozen pastries?

2. A school has a fundraiser. Students earn prize points based on the number of items they sell. The graph shows the ratio relationship between the number of items a student sells and the number of prize points the student earns.

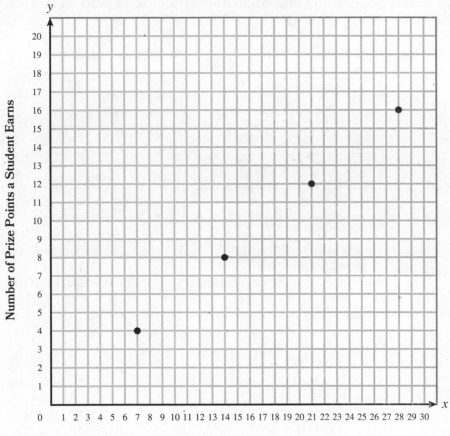

School Fundraiser

a. Fill in the blanks to make a true statement.

For every _____ more items a student sells, the student earns _____ more prize points.

b. How many prize points does a student earn if he sells 35 items?

c. The top seller earned 24 prize points. How many items did she sell?

3. A homemade modeling clay recipe calls for salt and flour.

   a. Use addition patterns to complete the ratio table.

Number of Cups of Salt	Number of Cups of Flour
10	6
15	9

   b. What is a ratio of the number of cups of salt to the number of cups of flour?

   c. Describe the ratio relationship by using the addition patterns in the ratio table.

4. Consider the graph of the ratio relationship between the amount of money in dollars Tara saves and the amount of money in dollars Tara spends.

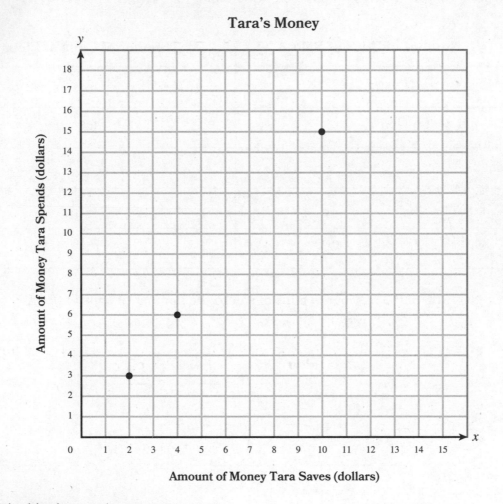

Tara's Money

a. Fill in the blank to make a true statement.

   The point (6, _____) lies on a line with the three given points.

b. Write the ordered pairs of two other points that lie on a line with the three given points.

c. Describe the ratio relationship by using the addition patterns in the graph.

5. The double number line shows the conversions between the number of cups and the number of quarts.

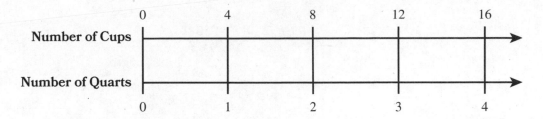

a. Fill in the blank to make a true statement.

There are _____ more cups for every 1 more quart.

b. Fill in the blank to make a true statement.

There are 8 more cups for every _____ more quarts.

6. The table shows the ratio relationship between the number of cups of yellow paint and the number of cups of blue paint needed to make batches of green paint. Complete the ratio table.

Number of Cups of Yellow Paint	Number of Cups of Blue Paint
3	
6	4
	6
15	

Remember

For problems 7 and 8, multiply.

7. $1,536 \times 42$

8. $2,347 \times 56$

9. Show that the ratio $2:5$ is equivalent to the ratio $14:35$.

10. Blake, Kayla, and Lacy form a relay team. They run equal distances in a 13-mile course. What distance in miles does each person run in the course? Choose all that apply.

A. $\frac{3}{13}$ miles

B. $\frac{13}{3}$ miles

C. $4\frac{1}{3}$ miles

D. $\frac{41}{3}$ miles

E. $\frac{4}{13}$ miles

F. $4\frac{1}{13}$ miles

Name

Date

Multiplication Patterns in Ratio Relationships

Making Concrete

1. Concrete is made by mixing sand, cement, and water. The graph shows the ratio relationship between the number of kilograms of sand and the number of kilograms of cement in a concrete recipe.

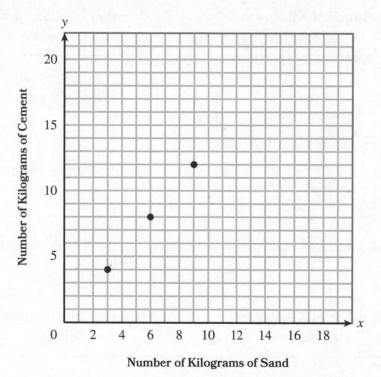

a. Write an ordered pair that represents one of the points plotted on the graph. Explain the meaning of the ordered pair in this situation.

So this explain the mix we need because for every 3 kg of cement we need 4 kg or sand

(3:4)

b. Write an ordered pair that belongs to this ratio relationship but does not represent a point plotted on the graph.

(12, 16)

c. Complete the ratio table. Include the numbers from the ordered pair you wrote in part (b) in the blank row.

Number of Kilograms of Sand	Number of Kilograms of Cement
3	4
6	8
9	12
12	16

d. Scott has 60 kilograms of sand to use for making concrete. How many kilograms of cement should he use?

Number of Kilograms of Sand	Number of Kilograms of Cement
3	4
6	8
9	12
60	80

Making Birdseed

2. Different mixes of birdseed attract different types of birds. Yuna's birdseed recipe calls for sunflower seeds and pumpkin seeds.

   a. Complete the table.

Number of Cups of Sunflower Seeds	Number of Cups of Pumpkin Seeds
3	
6	
9	6
18	
54	

   b. Use ratio language to describe the relationship between the number of cups of sunflower seeds and the number of cups of pumpkin seeds.

3. Blake's birdseed recipe calls for sunflower seeds and cracked corn. The double number line represents the relationship between the number of cups of sunflower seeds and the number of cups of cracked corn in his recipe. Use the double number line to determine the number of cups of cracked corn that Blake should use with 16 cups of sunflower seeds.

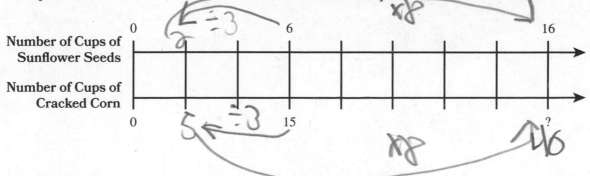

Making Sculptures

4. Eddie is making a sculpture of a person. He uses pipe cleaners for the arms and legs. The ratio of the sculpture's leg length in centimeters to its arm length in centimeters is $7:5$.

   a. If Eddie's sculpture has arms that are each 20 centimeters long, what is the length of each leg in centimeters?

   b. If Eddie's sculpture has legs that are each 63 centimeters long, what is the length of each arm in centimeters?

5. Jada is making a sculpture of a person. The ratio of her sculpture's leg length in inches to its arm length in inches is $8:6$. If the sculpture's arms are each 9 inches long, what is the length of each leg in inches?

Name _____ Date _____

A set of children's books has 5 pages of text for every 2 pages of illustrations. The table shows this ratio relationship.

a. Complete the ratio table.

Number of Pages of Text	Number of Pages of Illustrations
5	2
10	
15	
30	
55	

b. When there are 25 pages of text, how many pages of illustrations are there?

c. Explain how you used a multiplication pattern to find the solution to part (b).

b. In the coordinate plane provided, plot at least three points that each represent a possible ratio in this relationship.

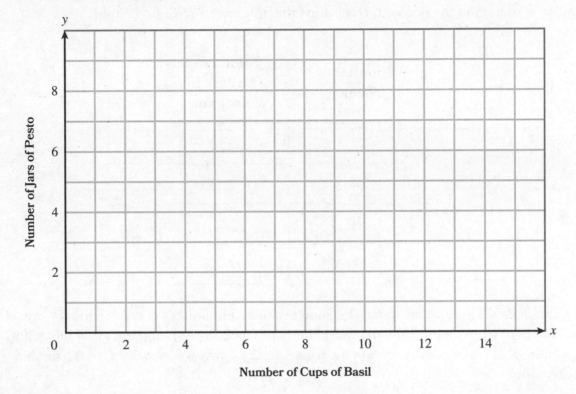

Number of Cups of Basil

c. Use the graph to determine the number of cups of basil the cook needs to make 4 jars of pesto. Use your solution to plot another point on the graph.

3. Riley saves money in a bank account to buy a used car. The ratio table shows the relationship between the number of weeks that have passed since Riley opened a bank account and the number of dollars in the account. Determine the unknown values in the table.

Number of Weeks	Number of Dollars in Account
2	350
4	700
10	
30	

4. An artist makes a gray wood stain. The stain requires 5 milliliters of black stain for every 3 milliliters of white paint. How many milliliters of black stain should the artist mix with 27 milliliters of white paint to create the same shade of gray wood stain? Use the double number line.

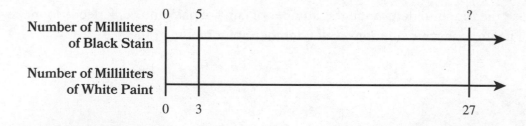

5. Scott pays $12 for 8 pounds of apples. What is the greatest number of pounds of apples Scott can buy with $60? Explain.

6. Kayla uses 4 pounds of shredded carrots and the juice of 2 lemons for a salad. How many lemons does Kayla need if she uses 6 pounds of carrots? Explain.

7. A taco spice recipe calls for 12 teaspoons of chili powder and 9 teaspoons of garlic salt. How many teaspoons of chili powder are needed to mix with 6 teaspoons of garlic salt? Explain.

Remember

For problems 8 and 9, multiply.

8. $1,773 \times 54$

9. $3,519 \times 26$

10. The table shows the ratio relationship between the number of minutes per day that Yuna reads and the number of minutes she spends doing math homework. Complete the ratio table.

Number of Minutes Yuna Reads	Number of Minutes Yuna Does Math Homework
18	6
27	9
36	
45	

11. A restaurant has 1 gallon of chicken soup and 3 quarts of vegetable soup. What is the total number of one-cup containers that this soup can fill?

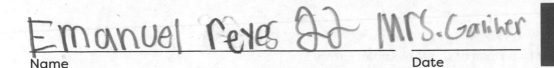

Emanuel reyes 22 Mrs. Gaither

Name   Date

Multiplicative Reasoning in Ratio Relationships

Following Recipes

1. A bread recipe calls for 2 cups of whole wheat flour for every 6 cups of white flour.

   a. Complete the ratio table.

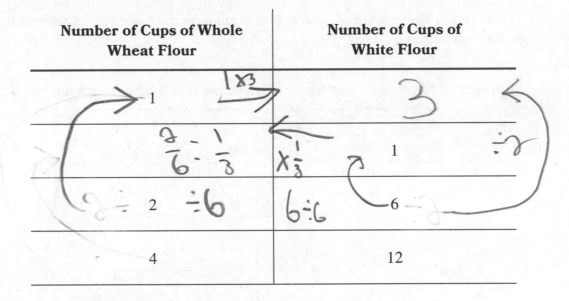

Number of Cups of Whole Wheat Flour	Number of Cups of White Flour
1	3
	1
2	6
4	12

   b. Complete the following statements.

   For every 1 cup of white flour, the recipe calls for _____ cup of whole wheat flour.

   For every 1 cup of whole wheat flour, the recipe calls for _____ cups of white flour.

   c. How are the first two rows of the table similar?

d. How are the first two rows of the table different?

e. The number of cups of white flour is always _____ times the number of cups of whole wheat flour.

f. The number of cups of whole wheat flour is always _____ times the number of cups of white flour.

2. The ratio table shows the number of cups of brown sugar and the number of cups of ketchup in a recipe for a homemade sauce.

Number of Cups of Brown Sugar	Number of Cups of Ketchup
	1
1	
2	3
4	6
6	9

a. Complete the ratio table.

b. Describe the relationship between the number of cups of brown sugar and the number of cups of ketchup.

c. The number of cups of ketchup is always _____ times the number of cups of brown sugar.

d. How many cups of ketchup need to be mixed with 10 cups of brown sugar to make the homemade sauce?

For problems 3–5, use the given ratio to complete the sentences.

3. Sana uses shaving cream and glue to make slime. The ratio of the number of tablespoons of shaving cream to the number of tablespoons of glue is 2 to 1. For every 1 tablespoon of glue Sana uses, she uses _____ tablespoon(s) of shaving cream. For every 1 tablespoon of shaving cream Sana uses, she uses _____ tablespoon(s) of glue.

4. Tyler fills fruit baskets with apples and bananas. The ratio of the number of apples to the number of bananas in each basket is 12 to 3. For every 1 apple in the basket, the basket has _____ banana(s). For every 1 banana in the basket, the basket has _____ apple(s).

5. In a recipe for trail mix, the ratio of the number of cups of cashews to the number of cups of almonds is 2 to 3. There are _____ times as many cups of almonds as cups of cashews. There are _____ times as many cups of cashews as cups of almonds.

Ratio Stations

Directions: Complete as many stations as you can in the time allotted. Create a ratio table or other diagram to explain your solution to the problem at each station.

Station 1	Station 2
Station 3	**Station 4**

Name _____ Date _____

Mr. Evans uses a ratio table to keep track of the numbers of cups of water and juice he drinks. He drinks 5 cups of water for every 2 cups of juice. Complete the ratio table.

Number of Cups of Water	5	10	15	1	
Number of Cups of Juice	2	4	6		1

a. How many cups of juice does Mr. Evans drink for every 1 cup of water he drinks?

b. How many cups of water does Mr. Evans drink for every 1 cup of juice he drinks?

7. Leo needs plaster to create masks for a costume party. To make the plaster, he mixes 2 cups of flour with 7 cups of water.

   a. How many cups of water does Leo use for every 1 cup of flour that he uses?

   b. How many cups of flour does Leo use for every 1 cup of water that he uses?

   c. If Leo uses 9 cups of flour, how many cups of water should he use?

8. Sasha makes coffee using 3 parts water with 2 parts espresso. Which table correctly shows the ratio relationship between the number of parts water and the number of parts espresso that Sasha uses?

A.

Number of Parts Water	Number of Parts Espresso
1	$\frac{3}{2}$
$\frac{2}{3}$	1
2	3

B.

Number of Parts Water	Number of Parts Espresso
1	$\frac{2}{3}$
$\frac{3}{2}$	1
3	2

C.

Number of Parts Water	Number of Parts Espresso
1	$\frac{2}{3}$
$\frac{3}{2}$	1
2	3

D.

Number of Parts Water	Number of Parts Espresso
1	$\frac{3}{2}$
$\frac{2}{3}$	1
3	2

9. Noah is making a shade of orange paint. He mixes 1 gallon of red paint with every 3 gallons of yellow paint. Based on this ratio, which statements are true? Choose all that apply.

   A. Noah mixes 1 gallon of yellow paint with every $\frac{1}{3}$ gallon of red paint.

   B. Noah mixes 2 gallons of red paint with every 6 gallons of yellow paint.

   C. There is 1 gallon of red paint in a 4-gallon mix of orange paint.

   D. There are 2 gallons of yellow paint in an 8-gallon mix of orange paint.

   E. A 4-gallon mix of orange paint would be $\frac{3}{4}$ red paint.

   F. A 4-gallon mix of orange paint would be $\frac{3}{4}$ yellow paint.

Remember

10. Multiply.

    $14{,}925 \times 36$

11. Ryan runs and walks every day. For every 4 minutes that he runs, he walks for 1 minute.

    a. Complete the ratio table.

Number of Minutes Ryan Runs	Number of Minutes Ryan Walks	Ratio	Ordered Pair
4	1	4 : 1	(4, 1)
8	2	8 : 2	(8, 2)
12			
16			

b. Label the axes on the coordinate grid. Then use the ordered pairs from the table to graph the ratio relationship.

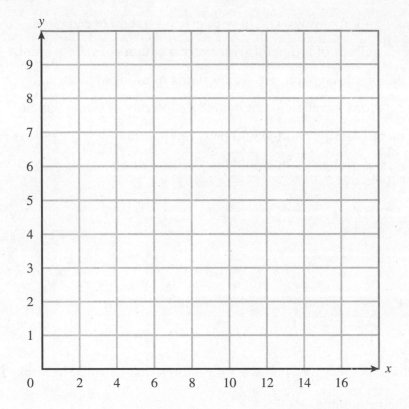

12. Lisa wants to outline her rectangular poster with duct tape. She needs two pieces of duct tape that each measure 54 centimeters and two pieces that each measure 72 centimeters. Lisa has 2.5 meters of duct tape. Does she have enough duct tape to outline the poster?

Name Emanuel reyes 22 mrs. G   **Date**

Applications of Ratio Reasoning

Multiple Ratios and Changing Ratios

1. Kelly has two gardens that each have an area of 40 square feet. He plants tomato seeds in one garden and green bean seeds in the other. He plants 1 tomato seed for every 8 square feet of garden area. He plants 1 green bean seed for every 2 square feet of garden area. Kelly plants the greatest number of seeds he can in each garden.

   a. What is the total number of tomato seeds Kelly plants?

   b. What is the total number of green bean seeds Kelly plants?

   c. What is the ratio of the total number of tomato seeds to the total number of green bean seeds Kelly plants?

   5:20

2. The art teacher's favorite paint color is a mixture of yellow, blue, and white paints. The mixture has 2 parts yellow paint for every 3 parts blue paint. It has 3 parts blue paint for every 5 parts white paint. If the teacher uses 8 jars of yellow paint, how many jars of white paint does he use?

3. The ratio of the number of stickers in bag A to the number of stickers in bag B is 4 : 3. Half of the stickers in bag A are moved to bag B. What is the new ratio of the number of stickers in bag A to the number of stickers in bag B?

4. Noah mixes 7 parts yellow paint for every 3 parts blue paint to make green paint. He adds 12 pints of blue paint to the mixture. Now the number of parts yellow paint is equal to the number of parts blue paint. How many pints of green paint did Noah have before adding more blue paint?

Changing Ratios Stations Activity

Start at the station indicated by your teacher. Move in order of station number. Your goal is to complete at least three stations in the time allotted. Draw diagrams or tables to support your thinking.

Station 1

Station 2

Station 3

Station 4

Station 5

Station 6

Names _____   Date _____

1. The ratio of the number of Sana's marbles to the number of Tyler's marbles is 1:8. After Tyler gives 9 marbles to Sana, the ratio of the number of Sana's marbles to the number of Tyler's marbles is 4:5.

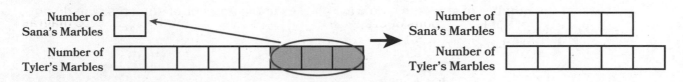

a. What is the total number of marbles that are represented by the three shaded units on the tape diagram?

b. What does each unit of the tape diagrams represent? Complete the tape diagrams by filling in the value of each unit in both tape diagrams.

c. What is the total number of marbles Sana and Tyler have?

2. In an art class, the ratio of the number of students who wear glasses to the total number of students is 2:7.

a. What is the ratio of the number of students who wear glasses to the number of students who do not wear glasses?

b. There are 8 students who wear glasses. What is the total number of students in the class?

c. Two students in the class who did not wear glasses now wear glasses. What is the new ratio of the number of students who wear glasses to the number of students who do not wear glasses?

3. An animal shelter with only cats and dogs has 40 animals. The ratio of the number of cats to the number of dogs at the shelter is $3:5$. If 5 more dogs arrive at the shelter, what is the new ratio of the number of cats to the number of dogs at the shelter? Explain.

4. Ryan mixes 3 parts white paint with 8 parts blue paint to make light blue paint. He adds a total of 10 tablespoons of white paint. His mixture is now equal parts white and blue paint. After adding the 10 tablespoons of white paint, what is the total number of tablespoons of paint Ryan has? Draw tape diagrams to show your thinking.

5. The sixth grade has two classrooms. Classroom A has 28 students. The ratio of the number of right-handed students to the number of left-handed students in classroom A is $5:2$. Classroom B has 27 students. The ratio of the number of right-handed students to the number of left-handed students in classroom B is $8:1$. What is the ratio of the total number of right-handed students to the total number of left-handed students? Explain.

6. Blake makes slime using glue and laundry soap. His recipe calls for 5 ounces of glue for every 2 ounces of laundry soap. Blake accidentally mixes 2 ounces of glue with 5 ounces of laundry soap. How many ounces of glue and how many ounces of laundry soap does Blake need to add so that his slime follows the original recipe? Explain.

7. A florist has two bouquets that each have the same number of flowers. She moves 4 flowers from bouquet A to bouquet B. The ratio of the number of flowers in bouquet A to the number of flowers in bouquet B is now $1:3$. How many flowers are now in each bouquet? Explain, or draw a tape diagram to show your thinking.

Remember

8. Multiply.

   $23,374 \times 58$

9. Jada uses beads to make necklaces. The ratio table shows the relationship between the number of red beads and the number of silver beads that Jada uses.

   a. Complete the ratio table.

Number of Red Beads	Number of Silver Beads
2	3
4	
6	
16	
30	

   b. Describe a multiplication pattern in the ratio table.

10. Find the quotient.

    $4,872 \div 12$

    A. 46

    B. 406

    C. 460

    D. 4,006

Spice Ratios

Some people love spice. Some people can't stand it. But no matter which group you belong to, we can all agree: What matters is not the *amount* of spice but its *ratio* to other ingredients.

A large amount of hot pepper spread across a huge amount of food won't taste very spicy.

And a small amount of hot pepper concentrated in a tiny amount of food may overwhelm you!

Overcoming a spicy feeling is tricky! Drinking cold water doesn't actually relieve the hot feeling. Instead, it's helpful to have milk, bread, or rice. That's because the chemicals that cause the spicy feeling attach to these other foods, meaning that they attach less to the receptors in your tongue.

(How much milk to drink? It depends how much spicy food you ate! Ratios never stop mattering.)

Name _____ Date _____

Multiple Ratio Relationships

Are They the Same Shade?

Choosing a Representation

Name _____ Date _____

The table shows the total cost of apples at store A. The graph shows the total cost of apples at store B. Do the table and the graph represent the same ratio relationship? Explain how you know.

Store A

Number of Pounds of Apples	Total Cost (dollars)
3	5
6	10
9	15
12	20
15	25

Store B

Number of Pounds of Apples

Name _____ Date _____

1. Smoothie store A and smoothie store B both make strawberry-banana smoothies. The graph shows the ratio relationship between the number of cups of strawberries and the number of bananas in each store's strawberry-banana smoothie.

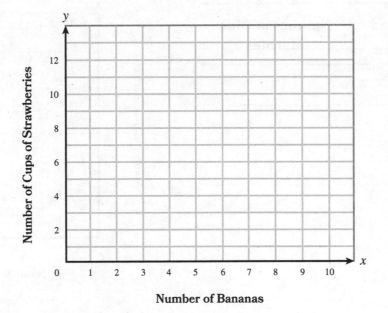

Number of Bananas

Should the smoothies have the same flavor? Explain.

2. Miss Baker gives each student in her class a bag that has red marbles and blue marbles in it. She asks each student to represent the ratio relationship between the number of red marbles and the number of blue marbles in their bag by using either a table or a graph. Is the ratio of the number of red marbles to the number of blue marbles in Julie's bag equivalent to the ratio of the number of red marbles to the number of blue marbles in Tara's bag? Explain how you know.

Julie's Marbles

Number of Red Marbles	Number of Blue Marbles
3	7
6	14
9	21
12	28

Tara's Marbles

Number of Blue Marbles (y-axis)

Number of Red Marbles (x-axis)

3. Scott and Yuna are mixing fabric dyes. They each create their favorite shade of red dye by mixing scarlet red dye and cherry red dye. Do Scott and Yuna create the same shade of red dye? If yes, what is the ratio of the number of ounces of scarlet red dye to the number of ounces of cherry red dye they both use? If not, what are the ratios that each person uses?

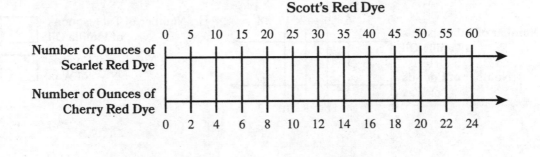

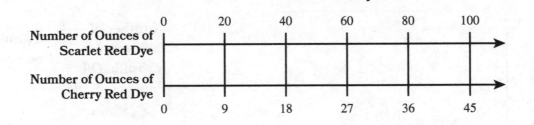

4. The sixth grade art class makes candles. Students use 3 tablespoons of vanilla oil for every 2 pounds of wax to make the candles. Which representations correctly show the ratio relationship between the number of tablespoons of vanilla oil and the number of pounds of wax? Choose all that apply.

A.
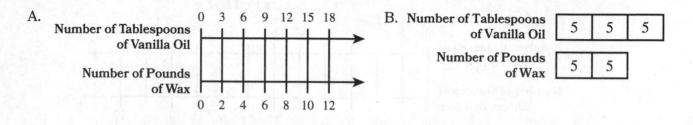

Number of Tablespoons of Vanilla Oil

Number of Pounds of Wax

B.
Number of Tablespoons of Vanilla Oil	5	5	5

Number of Pounds of Wax	5	5

C.
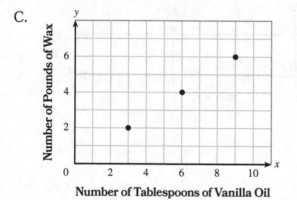

D.
Number of Tablespoons of Vanilla Oil	Number of Pounds of Wax
3	2
4	3
5	4

E.
Number of Tablespoons of Vanilla Oil	Number of Pounds of Wax
6	4
12	8
18	12

F.
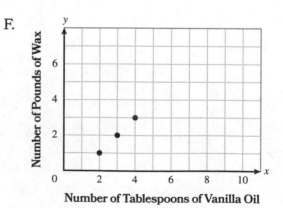

5. Adesh's recipe for tea calls for 4 tablespoons of milk in 3 cups of tea. Kelly's recipe for tea calls for 7 tablespoons of milk in 5 cups of tea.

   a. Show the ratio relationship between the number of tablespoons of milk and the number of cups of tea in Adesh's recipe. Create a graph, double number line, or table.

   b. Show the ratio relationship between the number of tablespoons of milk and the number of cups of tea in Kelly's recipe. Create a graph, double number line, or table.

   c. Do Adesh's recipe and Kelly's recipe use the same ratio relationship between the number of tablespoons of milk and the number of cups of tea? Explain.

Remember

For problems 6 and 7, divide.

6. $8,448 \div 2$

7. $1,272 \div 3$

8. The table shows the ratio relationship between the number of ounces of raisins and the number of ounces of nuts in a trail mix recipe.

   a. Complete the table.

Number of Ounces of Raisins	Number of Ounces of Nuts
3	2
6	
9	
1	
	1

   b. How many ounces of raisins are needed for every 1 ounce of nuts?

   c. How many ounces of nuts are needed for every 1 ounce of raisins?

9. Which pair of statements accurately describes the ratio relationship shown in the table?

Number of Cups of Flour	Number of Cups of Water
3	1
6	2
9	3
12	4

A. The ratio of the number of cups of flour to the number of cups of water is $1:2$. For every 1 cup of flour, there are 2 cups of water.

B. The ratio of the number of cups of flour to the number of cups of water is $1:3$. For every 1 cup of flour, there are 3 cups of water.

C. The ratio of the number of cups of flour to the number of cups of water is $3:1$. For every 3 cups of flour, there is 1 cup of water.

D. The ratio of the number of cups of flour to the number of cups of water is $3:6$. For every 3 cups of flour, there are 6 cups of water.

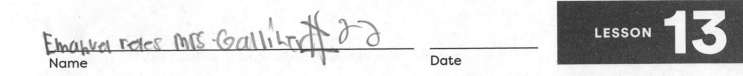

LESSON 13

Name _Emanuel rates mrs Gallibert J J_ Date _____

Comparing Ratio Relationships, Part 1

Using Ratio Tables to Compare Two Ratio Relationships

1. Yuna and Tyler make lemonade. The tables show the ratio relationship between the number of cups of water and the number of tablespoons of lemon juice concentrate in each recipe.

Yuna's Recipe

Number of Cups of Water	Number of Tablespoons of Lemon Juice Concentrate
2	6
3	9
5	15
7	21

Tyler's Recipe

Number of Cups of Water	Number of Tablespoons of Lemon Juice Concentrate
1	4
2	8
4	16
10	40

a. What is a ratio of the number of cups of water to the number of tablespoons of lemon juice concentrate in Yuna's recipe? 2:6, 3:9, 5:15, 7:21

b. What is a ratio of the number of cups of water to the number of tablespoons of lemon juice concentrate in Tyler's recipe? 1:4, 2:8, 4:16, 10:40

c. Based on the tables, whose lemonade should have a stronger lemon flavor? How do you know?

Tylers recipe is going to have as stroger for ever 2 cups of water te and 6 the spoons of lemon juice lemon juice becaus

2. Store A and store B both sell oranges. The tables show the total cost in dollars and the number of pounds of oranges at each store.

Store A			Store B	
Number of Pounds of Oranges	**Total Cost (dollars)**		**Number of Pounds of Oranges**	**Total Cost (dollars)**
4	6		5	9
8	12		10	18
12	18		15	27
16	24		20	36

a. Does the table for store A represent a ratio relationship? Explain.

b. Does the table for store B represent a ratio relationship? Explain.

c. Do the two tables represent the same ratio relationship? Explain.

d. Which store charges more per 1 pound of oranges? Explain.

3. Lisa and Tara run laps at soccer practice. Both girls record the number of minutes they run laps in ratio tables.

Lisa	Number of Minutes	4	12	20	28
	Number of Laps	2	6	10	14

Tara	Number of Minutes	2	8	14	20
	Number of Laps	1	4	7	10

Who runs faster? Explain.

Using Parts and Wholes to Compare Two Ratio Relationships

4. Mr. Sharma's class mixes dye to make tie-dye shirts. The class mixes yellow dye and red dye to make two different mixtures of orange dye. Which mixture should appear more yellow? Use the information in the tables to support your answer. Circle the table rows used to determine your answer.

Mixture A

Number of Parts Yellow Dye	Number of Parts Red Dye	Total Number of Parts
3	5	8
6	10	16
9	15	24
12	20	32
15	25	40
18	30	48
21	35	56

Mixture B

Number of Parts Yellow Dye	Number of Parts Red Dye	Total Number of Parts
5	7	12
10	14	24
15	21	36
20	28	48
25	35	60
30	42	72
35	49	84

Using Ratio Tables to Compare Three Ratio Relationships

5. Lacy, Riley, and Adesh each make orange and vanilla ice pops by using the same two ingredients. The tables show the number of cups of orange juice and the number of cups of vanilla yogurt used by each person.

Lacy's Ice Pops		Riley's Ice Pops		Adesh's Ice Pops	
Number of Cups of Orange Juice	Number of Cups of Vanilla Yogurt	Number of Cups of Orange Juice	Number of Cups of Vanilla Yogurt	Number of Cups of Orange Juice	Number of Cups of Vanilla Yogurt
10	12	4	6	3	4
20	24	8	12	6	8
30	36	12	18	9	12

Based on the tables, order the ice pops from the one that should have the strongest orange flavor to the one that should have the weakest orange flavor. Explain how you used the numbers in the tables to determine the order.

EXIT TICKET 13

Name _____   Date _____

Leo and Ryan each make salsa. The ratio tables show the relationship between the number of cups of tomatoes and the number of cups of spicy peppers in each of their salsas.

Leo's Salsa		Ryan's Salsa	
Number of Cups of Tomatoes	Number of Cups of Spicy Peppers	Number of Cups of Tomatoes	Number of Cups of Spicy Peppers
9	2	20	4
18	4	25	5
27	6	30	6
36	8	35	7

Based on the tables, whose salsa should be spicier? Explain how you know.

Name

Date

1. The graph shows the ratio relationship between the number of lemons and the number of cups of water in the recipe for lemonade A. The recipe for lemonade B calls for 1 lemon for every $2\frac{1}{2}$ cups of water.

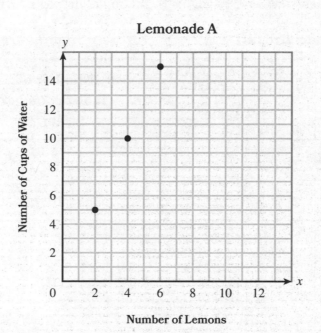

Lemonade A

a. Complete the ratio table for lemonade B.

Lemonade B

Number of Lemons	Number of Cups of Water
1	
2	
3	
4	

b. Based on the graph and the table, should one lemonade have a stronger lemon flavor than the other? Explain how you know.

2. Tara and Noah each make strawberry yogurt by mixing plain yogurt with strawberry jam.

Tara's Strawberry Yogurt

Number of Tablespoons of Jam	Number of Cups of Plain Yogurt
4	2
8	4
12	6
16	8
20	10

Noah's Strawberry Yogurt

Number of Tablespoons of Jam	Number of Cups of Plain Yogurt
5	3
10	6
15	9
20	12
25	15

a. Describe the ratio relationship between the number of tablespoons of jam and the number of cups of plain yogurt in Tara's strawberry yogurt.

b. Describe the ratio relationship between the number of tablespoons of jam and the number of cups of plain yogurt in Noah's strawberry yogurt.

c. Based on the tables, whose strawberry yogurt should have a stronger strawberry flavor? Explain.

3. Cans A and B are filled with pink paint. The ratio tables show the relationship between the number of parts white paint and the number of parts red paint in each can.

Can A

Number of Parts White Paint	Number of Parts Red Paint
7	4
14	8
21	12
28	16
35	20

Can B

Number of Parts White Paint	Number of Parts Red Paint
8	5
16	10
24	15
32	20
40	25

Based on the ratio tables, which can is filled with paint that should look redder? Explain.

4. The ratio tables show the relationship between the number of ounces of coconut oil and the number of ounces of olive oil in two different soaps.

   a. Complete both ratio tables.

Soap 1

Number of Ounces of Coconut Oil	Number of Ounces of Olive Oil	Total Number of Ounces of Oil
3	4	7
9		21
	16	28
30	40	

Soap 2

Number of Ounces of Coconut Oil	Number of Ounces of Olive Oil	Total Number of Ounces of Oil
4	6	
	12	20
12	18	30
28		70

   b. What is the ratio of the number of ounces of coconut oil to the number of ounces of olive oil in soap 1?

   c. What is the ratio of the number of ounces of coconut oil to the number of ounces of olive oil in soap 2?

   d. Coconut oil produces more bubbles in soap than olive oil. Based on the ratio tables, which soap should have more bubbles? Explain.

e. Adesh reasons that soap 1 should have more bubbles than soap 2. For every 12 ounces of olive oil, soap 1 has 9 ounces of coconut oil while soap 2 only has 8 ounces of coconut oil.

Sana reasons that soap 1 should have more bubbles than soap 2. For every 70 ounces of oil, soap 1 has 30 ounces of coconut oil while soap 2 only has 28 ounces of coconut oil.

Which statement is true?

A. Only Adesh's reasoning is correct.

B. Only Sana's reasoning is correct.

C. Both Adesh's and Sana's reasoning are correct.

D. Neither Adesh's nor Sana's reasoning is correct.

Remember

For problems 5 and 6, divide.

5. $2,405 \div 5$

6. $9,642 \div 3$

7. Kayla and Yuna have the same amount of money. After Kayla spends $24.00, the ratio of the amount of money in dollars Kayla has to the amount of money in dollars Yuna has is $5 : 8$. How much money does Kayla have now?

For problems 8 and 9, complete the tables by converting each measurement to the given unit.

8.

Number of Minutes	Number of Seconds
8	
10	

9.

Number of Feet	Number of Inches
2	
8	

Name Emanuela Kela Galiher 22   Date

Comparing Ratio Relationships, Part 2

Flour Mixtures

1. Two restaurants use different flour mixtures in their pancake batter recipes. The ratio tables show the relationship between the number of cups of flour and the number of tablespoons of salt in the mixtures.

Restaurant A		Restaurant B	
Number of Cups of Flour	**Number of Tablespoons of Salt**	**Number of Cups of Flour**	**Number of Tablespoons of Salt**
27	18	22	11
30	20	24	12
33	22	26	13

Which restaurant's flour mixture is saltier? Explain.

reastarant A is Saltier because I got both to 28 cups of flour and reastarant A was saltier.

2. Beekeepers add sugar water to the diet of honeybees. In the spring, the sugar water mixture helps promote colony growth. In the fall, the sugar water mixture helps the bees survive. The ratio tables show the number of cups of water and the number of cups of sugar in the spring sugar water mixture and in the fall sugar water mixture.

Spring Sugar Water Mixture

Number of Cups of Sugar	Number of Cups of Water
15	10
18	12
27	18

×7   105   70

Fall Sugar Water Mixture

Number of Cups of Sugar	Number of Cups of Water
14	7
32	16
42	21

×7   ×10   146   70

Based on the tables, which sugar water mixture is sweeter? Explain.

Tiles

3. A flooring company offers two different designs that each use white square tiles and blue square tiles. All of the tiles are the same size. The ratio tables show the number of white tiles and the number of blue tiles needed for each design.

Floor Design A

Number of White Tiles	Number of Blue Tiles
20	30
40	60

Floor Design B

Number of White Tiles	Number of Blue Tiles
15	35
45	105

Based on the tables, which floor design should appear bluer? Explain.

4. The same flooring company offers two different designs that each use gold square tiles and black square tiles. All of the tiles are the same size. The ratio tables show the number of gold tiles and number of black tiles needed for each design.

Floor Design Y

Number of Gold Tiles	Number of Black Tiles
13	52
18	72

Floor Design Z

Number of Gold Tiles	Number of Black Tiles
16	48
20	60

Based on the tables, which floor design should appear more gold? Explain.

Name _____   Date _____

Café A and café B each have a hot cocoa recipe. The ratio tables show the relationship between the number of tablespoons of cocoa powder and the number of ounces of milk in each café's recipe.

Café A	
Number of Tablespoons of Cocoa Powder	Number of Ounces of Milk
2	9
4	18
6	27

Café B	
Number of Tablespoons of Cocoa Powder	Number of Ounces of Milk
3	11
9	33
15	55

Based on the tables, which café's hot cocoa should have a stronger cocoa flavor? How did you determine your answer?

3. The tables show the ratio relationship between the number of cups of sugar and the number of lemons for Lisa's and Yuna's lemonade recipes.

<table>
<tr><th colspan="2">Lisa's Recipe</th><th colspan="2">Yuna's Recipe</th></tr>
<tr><th>Number of Cups of Sugar</th><th>Number of Lemons</th><th>Number of Cups of Sugar</th><th>Number of Lemons</th></tr>
<tr><td>1</td><td>2</td><td>4</td><td>5</td></tr>
<tr><td>2</td><td>4</td><td>5</td><td>6.25</td></tr>
<tr><td>3</td><td>6</td><td>6</td><td>7.5</td></tr>
</table>

a. Based on the tables, whose lemonade should have a stronger lemon flavor? Explain how you know.

b. Each girl has exactly 3 lemons to make lemonade. How many cups of sugar does each girl need?

4. Riley mixes paint to paint pumpkins. The ratio table shows the number of drops of blue paint and the number of drops of white paint that Riley mixes.

Riley's Pumpkin Paint

Number of Drops of Blue Paint	Number of Drops of White Paint
15	3
60	12
75	15

a. If Riley uses 1 drop of white paint, how many drops of blue paint does she use?

b. Write a ratio of the number of drops of blue paint to the number of drops of white paint that would make a darker shade of blue than Riley's paint.

c. Write a ratio of the number of drops of blue paint to the number of drops of white paint that would make a lighter shade of blue than Riley's paint.

Remember

For problems 5 and 6, divide.

5. $3{,}744 \div 6$

6. $7{,}389 \div 9$

7. The table shows the number of minutes and the number of miles that Toby rides his bike. The graph shows the number of minutes and the number of miles that Kelly rides his bike. Do the table and the graph represent the same ratio relationship? Explain how you know.

Toby's Bike Ride

Number of Minutes	Number of Miles
15	3
30	6
60	12
90	18

Kelly's Bike Ride

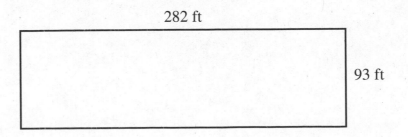

8. Find the area of Tyler's rectangular garden in square yards. (1 yard = 3 feet)

282 ft

93 ft

Name _Emanvel reles#22   Gallihar_

Date _____

The Value of the Ratio

1. Both cereal A and cereal B are made with marshmallows and oats. The recipe for one batch of cereal A calls for 12 pounds of marshmallows and 48 pounds of oats. The recipe for one batch of cereal B calls for 16 pounds of marshmallows and 80 pounds of oats.

   a. Complete the ratio tables.

   Cereal A

Number of Pounds of Marshmallows	Number of Pounds of Oats
1	4
$\frac{1}{4}$	1
12	48

   Cereal B

Number of Pounds of Marshmallows	Number of Pounds of Oats
1	5
$\frac{1}{5}$	1
16	80

   b. If you prefer more marshmallows in your cereal, which cereal would you choose? Explain.

   If I prefer marshmalloes.. in my ceral I woud rather Ceral A.I would rather ceral A because according to my math I is more than $\frac{1}{5}$. Ceral B has more oats $\frac{1}{4}$ is more so ceras has more marshmallow. ceral A has less

Cereal

For problems 2 and 3, fill in the blank with the correct value.

2. For the cereal A recipe, the number of pounds of marshmallows is ___$\frac{1}{4}$___ times the number of pounds of oats.

3. For the cereal B recipe, the number of pounds of marshmallows is ___$\frac{1}{5}$___ times the number of pounds of oats.

Fraction (A)

In conclusion ceral A has more marshmallow.

For problems 4 and 5, fill in the blank with the correct value.

4. For the cereal A recipe, the number of pounds of oats is ___1.5___ times the number of pounds of marshmallows.

5. For the cereal B recipe, the number of pounds of oats is _____ times the number of pounds of marshmallows.

Favorite Blue

6. Toby and Tara each paint their rooms their favorite shade of blue. Toby uses a paint mixture of 1 pint of blue paint for every 7 pints of white paint. Tara uses a paint mixture of 2 pints of blue paint for every 12 pints of white paint.

   a. What is a ratio of the number of pints of blue paint to the number of pints of white paint that Toby uses?

   b. What is the value of the ratio that compares the number of pints of blue paint to the number of pints of white paint in Toby's paint mixture?

   c. Describe what the value of the ratio in part (b) represents in this situation.

   d. What is a ratio of the number of pints of blue paint to the number of pints of white paint that Tara uses?

e. What is the ratio of the number of pints of blue paint to the number of pints of white paint that Tara uses when the number of pints of blue paint is 1?

f. What is the value of the ratio that compares the number of pints of blue paint to the number of pints of white paint in Tara's paint mixture?

g. Describe what the value of the ratio in part (e) represents in this situation.

h. Whose shade of blue is lighter? Explain.

i. Yuna also paints her room her favorite shade of blue. Yuna claims that her paint mixture is a darker shade of blue than both Toby's and Tara's paint mixtures. Write a ratio of the number of pints of blue paint to the number of pints of white paint that could represent Yuna's paint mixture. Explain.

7. The table shows ratios that each compare the number of pints of blue paint to the number of pints of white paint in different paint mixtures. Decide whether each ratio represents a paint mixture that is a lighter or darker shade of blue than Toby's paint mixture.

Ratio of Number of Pints Blue Paint to Number of Pints White Paint	Lighter than Toby's Paint	Darker than Toby's Paint
1 : 2		
1 : 8		
2 : 10		
2 : 20		
3 : 15		

Name _____   Date _____

Lacy and Sasha make apple pies. Lacy uses 15 cups of apples for every 2 pies. Sasha uses 18 cups of apples for every 3 pies. Who uses fewer cups of apples per pie? How do you know?

Name _____   Date _____

1. Blake works 4 hours and earns a total of $44.00. Leo works 7 hours and earns a total of $84.00.

   a. Blake earns _____ for every 1 hour that he works.

   b. Leo earns _____ for every 1 hour that he works.

   c. Leo and Blake each work one 8-hour shift this weekend. Who earns more money? Explain how you know.

2. Julie buys 5 tickets to a concert for a total of $110.00. Yuna buys 4 tickets to a different concert for a total of $84.00.

   a. What is a ratio that relates the amount of money in dollars Julie pays to the number of concert tickets she buys?

   b. What is the value of the ratio that you wrote in part (a)?

   c. Describe what the value of the ratio from part (b) represents in this situation.

d. What is the value of the ratio that describes the amount of money in dollars Yuna pays for 1 concert ticket?

e. Julie and Yuna each want to purchase 2 more concert tickets at the same price per ticket they paid before. Who will pay less for 2 more concert tickets?

3. Noah's snow cone recipe calls for 6 tablespoons of syrup for every 12 ounces of crushed ice. Kayla's snow cone recipe calls for 8 tablespoons of syrup for every 24 ounces of crushed ice.

a. Write a ratio that relates the number of tablespoons of syrup to the number of ounces of crushed ice in Noah's snow cone recipe.

b. What is the value of the ratio that you wrote in part (a)?

c. Describe what the value of the ratio from part (b) represents in this situation.

d. What is the value of the ratio that describes the number of tablespoons of syrup used for every 1 ounce of crushed ice in Kayla's snow cone recipe?

e. Whose snow cone recipe makes a snow cone with a stronger flavor? Explain.

4. Toby runs 5 laps in 10 minutes during track practice. Ryan runs 4 laps in 6 minutes.

   a. Write a ratio that relates the number of minutes Toby runs to the number of laps he runs.

   b. Write a ratio that relates the number of minutes Ryan runs to the number of laps he runs.

   c. Who runs at a faster pace? Use the value of the ratio to explain how you know.

5. Blake and Eddie each make strawberry lemonade. Blake uses 19 strawberries and makes 5 glasses of lemonade. Eddie uses 18 strawberries and makes 4 glasses of lemonade. Who uses more strawberries per glass of lemonade? Use the value of the ratio to explain how you know.

Remember

For problems 6 and 7, divide.

6. $3,010 \div 7$

7. $5,184 \div 6$

8. The ratio tables show the relationship between the number of cups of blueberries and the number of cups of yogurt in two different smoothies. Which smoothie should have a stronger blueberry flavor? Explain how you know.

Smoothie 1

Number of Cups of Blueberries	Number of Cups of Yogurt
2	5
4	10
6	15

Smoothie 2

Number of Cups of Blueberries	Number of Cups of Yogurt
3	6
6	12
9	18

For problems 9 and 10, complete the table by converting each measurement to the given unit.

9.

Number of Meters	Number of Centimeters
7	
10	
4.5	

10.

Number of Feet	Number of Yards
12	
9	
2	

The Six-Minute Mile

What's the fastest you can run?

It probably depends on the distance. For example, if you're running 50 or 100 meters, then you only need to maintain your top speed for less than 20 seconds. You can sprint as hard as you want! But if you're running several miles, you'll need to go much slower to conserve your energy.

The record pace for a marathon is roughly 13 miles per hour.

The record pace for a single mile is over 16 miles per hour.

And the record pace for a hundred meters is over ⬤ miles per hour!

Name _____ Date _____

Speed

1. Fill in the boxes of the bracket to show which animal wins each race.

Interpreting Speed

2. A squirrel runs at a speed of 12 miles per hour. Complete the table that shows the number of hours and the number of miles the squirrel runs at that speed.

Number of Hours	Number of Miles
1	12
2	24
3	36
4	48

3. A house mouse runs at a constant speed. It runs 4 miles per hour slower than the speed a squirrel runs.

    a. Determine the speed the house mouse runs in miles per hour. Use the speed the squirrel runs from problem 2.

    b. Interpret the meaning of the speed the house mouse runs.

4. When a peregrine falcon dives to catch its prey, it is the fastest animal in the world. The ratio table shows the number of seconds and the number of meters the peregrine falcon dives at a constant speed.

    a. Complete the ratio table.

Number of Seconds	Number of Meters
1	
2	176
3	264

    b. What is the speed the peregrine falcon dives in meters per second?

    c. Interpret the meaning of the speed the peregrine falcon dives.

5. The slowest animal in the world is a sloth. The double number line shows the number of meters and number of minutes a sloth crawls at a constant speed.

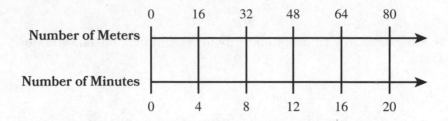

a. Determine the speed the sloth crawls in meters per minute.

b. Interpret the meaning of the speed the sloth crawls.

Speed, Distance, and Time

6. The double number line shows the number of miles and the number of hours a hummingbird flies.

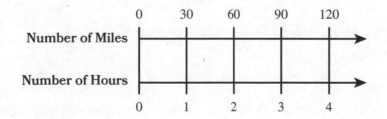

a. How many miles does the hummingbird fly in 5 hours?

b. How many hours does it take the hummingbird to fly 45 miles?

c. How many miles does the hummingbird fly in 30 minutes, or half an hour?

d. How many miles does the hummingbird fly in 1 minute, or $\frac{1}{60}$ of an hour?

7. A horsefly flies 70 kilometers in 30 minutes at a constant speed.

a. Create a double number line to determine the number of kilometers the horsefly flies in 150 minutes at this speed.

b. How many kilometers does the horsefly fly in 1 hour?

c. What is the speed the horsefly flies in kilometers per hour?

d. What is the speed the horsefly flies in kilometers per minute? Justify your reasoning.

Comparing Ratios and Rates

8. Ryan goes to a movie with his friends. He spends $24.00 to buy 4 buckets of popcorn.

    a. What is a ratio that relates the number of dollars Ryan spends to the number of buckets of popcorn he buys?

    b. What is the rate in dollars per bucket?

9. Lisa checks the nutrition facts on a bag of granola.

    a. There are 2 grams of protein for every 1 serving of granola. What is the rate in grams of protein per serving?

    b. The ratio of the number of grams of protein to the number of cups of granola is $32 : 4$. What is the rate in grams of protein per cup?

Name _____ Date _____

Riley runs at a constant speed of 6 miles per hour.

a. Interpret the meaning of Riley's speed.

b. Create a double number line to determine the number of miles Riley runs in 2 hours at this speed.

c. What amount of time does it take Riley to run 3 miles at this speed?

Name _____   Date _____

1. Karl Benz drove the first car in Mannheim, Germany, in 1886. The car traveled at a top speed of 10 miles per hour. Assume the car kept that constant speed.

   a. Interpret the meaning of the car's speed.

   b. Use your answer from part (a) to complete the ratio table.

Number of Hours	Number of Miles

   c. Create a double number line to represent this situation.

2. The ratio table shows the number of minutes and the number of meters a Galápagos tortoise walks.

Number of Minutes	Number of Meters
3	15
6	30
9	45

a. Determine the speed the tortoise walks in meters per minute.

b. At this speed, how many minutes does it take the tortoise to walk 25 meters?

c. At this speed, how many meters can the tortoise walk in 15 minutes?

3. A satellite in space travels at a constant rate. The double number line shows the number of kilometers and the number of hours a satellite travels.

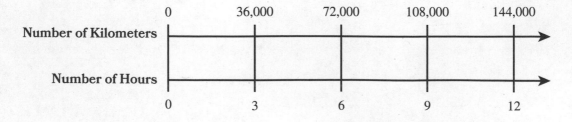

a. At this rate, how many kilometers does the satellite travel in 3 hours?

b. At this rate, how many kilometers does the satellite travel in 1 hour?

c. At this rate, how many kilometers does the satellite travel in half an hour?

d. At what speed does the satellite travel in kilometers per hour?

e. At this rate, how many hours does it take the satellite to travel 60,000 kilometers?

4. A bird named Zac the Macaw holds the world record for the most canned drinks opened in 1 minute by a parrot. Zac the Macaw opened 35 canned drinks in 1 minute.[1]

   a. Ryan says that at this rate, Zac the Macaw can open more than 100 canned drinks in 3 minutes. Do you agree with Ryan? Draw a double number line to support your answer.

   b. At this rate, how many canned drinks can Zac the Macaw open in 10 minutes?

5. The ratio of the number of syllables Lisa speaks to the number of seconds she speaks is 40 : 10. Assume that Lisa speaks at a constant rate.

   a. At what rate in syllables per second does Lisa speak?

   b. At this rate, how many syllables does Lisa speak in 30 seconds?

---

1 Guinness Book of World Records, "Most Canned Drinks Opened by a Parrot in One Minute."

6. Toby completes 10 homework problems in 5 minutes. Assume that Toby completes homework problems at a constant rate.

   a. What is a ratio that relates the number of homework problems Toby completes to the number of minutes?

   b. What is the rate in homework problems completed per minute?

7. One pint of frozen yogurt has 440 calories for 4 servings.

   a. What is a ratio that relates the number of calories to the number of servings in this pint of frozen yogurt?

   b. What is the rate in calories per serving?

8. Jada rakes leaves to earn extra money. The ratio of the number of hours she rakes leaves to the number of dollars she earns is $4:30$. What is Jada's rate in dollars per hour?

Remember

For problems 9 and 10, divide. Write the quotient and the remainder on separate lines.

9. $2,561 \div 3$

   Quotient: _____

   Remainder: _____

10. $5,218 \div 4$

   Quotient: _____

   Remainder: _____

11. Yuna and Scott use the same powdered lemonade mix to make lemonade. The tables show the ratio relationship between the number of tablespoons of lemonade mix and the number of ounces of water they each use. Based on the tables, whose lemonade should have a weaker lemon flavor? Explain how you know.

<table>
<thead>
<tr><th colspan="2">Yuna's Lemonade</th></tr>
<tr><th>Number of Tablespoons of Lemonade Mix</th><th>Number of Ounces of Water</th></tr>
</thead>
<tbody>
<tr><td>2</td><td>12</td></tr>
<tr><td>6</td><td>36</td></tr>
<tr><td>12</td><td>72</td></tr>
</tbody>
</table>

<table>
<thead>
<tr><th colspan="2">Scott's Lemonade</th></tr>
<tr><th>Number of Tablespoons of Lemonade Mix</th><th>Number of Ounces of Water</th></tr>
</thead>
<tbody>
<tr><td>3</td><td>15</td></tr>
<tr><td>5</td><td>25</td></tr>
<tr><td>10</td><td>50</td></tr>
</tbody>
</table>

For problems 12–16, round the number to the nearest tenth.

12. 23.38

13. 23.451

14. 23.309

15. 23.055

16. 22.962

Name _____　　　Date _____

Rates

The Unit Rate

1. The double number line represents the ratio relationship between the number of miles and the number of hours that Blake rides his bike.

Blake's Bike Rides

a. What is Blake's speed in miles per hour? Interpret its meaning.

Blake's speed in 1 hour is 15 milles.
Accooding to my math brake should
ride 15 miles in 1 hour.

b. At this speed, how many hours does it take Blake to ride 30 miles?

It would take brake 2 hours
to ride 30 milles. According to my
math it I add
it would give
give 3 miniles.

5 ˑ 75

c. At this speed, how many miles does Blake ride if he rides his bike for 5 hours?

2. The ratio table represents the relationship between the number of miles and the number of hours that Kelly rides her bike. Who rides faster, Kelly or Blake? Explain.

Blake's speed is 15 miles per hours.

Kelly's Bike Rides

Number of Miles	Number of Hours
64	4
128	8
192	12

3. The double number line represents the ratio relationship between the total cost in dollars and the number of gallons of gasoline.

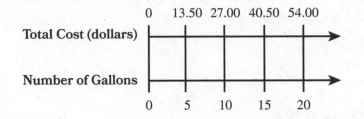

a. What is the rate in dollars per gallon? Interpret its meaning.

b. What is the unit rate?

c. At this rate, what is the total cost to fill a 26-gallon tank with gasoline if it is empty?

d. A driver spends a total of $24.30 on gasoline. At this rate, how many gallons of gasoline does the driver buy?

Another Unit Rate

4. Yuna types at a constant rate of 2 minutes per page. Her computer teacher, Miss Song, types at a constant rate of 2 pages per minute.

    a. Complete the ratio tables.

Yuna's Typing		Miss Song's Typing	
Number of Pages	**Number of Minutes**	**Number of Pages**	**Number of Minutes**
1	2	2	1

    b. How many pages can Yuna type in 50 minutes? 100 minutes?

    c. What is Yuna's rate in pages per minute? What is the unit rate?

    d. How many minutes does it take Miss Song to type 50 pages? 100 pages?

    e. What is Miss Song's rate in minutes per page? What is the unit rate?

Rates and the Coordinate Plane

5. The graph represents the ratio relationship between the number of cans of peas and the total cost in dollars of the cans of peas.

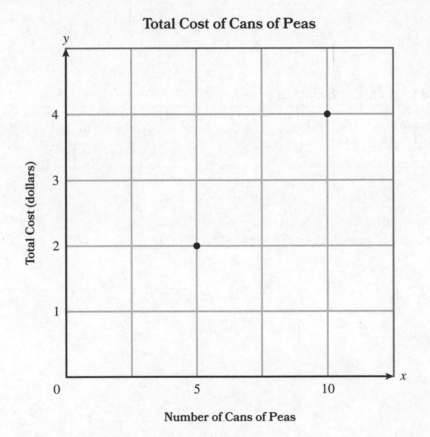

Total Cost of Cans of Peas

a. Write a ratio that relates the number of cans of peas to the total cost in dollars. What is the rate in cans of peas per dollar? What is the unit rate?

b. Write a ratio that relates the total cost in dollars to the number of cans of peas. What is the rate in dollars per can of peas? What is the unit rate?

c. What is the total cost of 8 cans of peas?

d. How many cans of peas can someone buy for $8.00?

6. Suppose that a professional auto racer drives one of the world's fastest sports cars at its top speed. The graph represents the ratio relationship between the number of hours and the number of miles the auto racer drives the sports car. Assume that the auto racer drives the sports car at a constant rate.

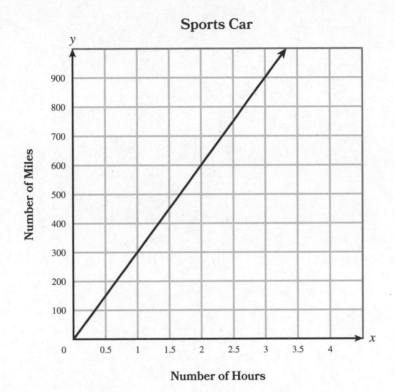

Sports Car

a. Based on the graph, what is the rate the auto racer drives the sports car in miles per hour? What is the unit rate?

b. Based on the graph, what is the rate the auto racer drives the sports car in hours per mile? What is the unit rate?

c. Suppose you could ride your bike for 12 hours at a constant speed of 15 miles per hour. How many hours would it take the auto racer to drive the sports car the same distance?

Name

Date

Sasha swims 4 laps in 2 minutes.

a. What is Sasha's rate in laps per minute? What is the unit rate?

b. What is Sasha's rate in minutes per lap? What is the unit rate?

Name _____ Date _____

1. Lacy earns a rate of $20.00 per hour tutoring.

   a. What is the unit rate?

   b. Complete the ratio table to show the number of hours Lacy tutors and the total amount of money she earns.

Number of Hours	Total Amount Earned

   c. At this rate, how much money does Lacy earn if she tutors for 8 hours?

   d. At this rate, how many hours must Lacy tutor to earn $200.00?

2. Adesh is jumping rope. The double number line represents the number of minutes and the number of jumps Adesh does.

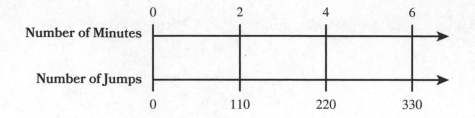

a. What is Adesh's jumping rate in jumps per minute?

b. What is the unit rate?

c. At this rate, how many jumps can Adesh do in 5 minutes?

3. Blake fills 10 water balloons in 4 minutes.

a. What is this rate in water balloons filled per minute? What is the unit rate?

b. What is this rate in minutes per water balloon filled? What is the unit rate?

4. A grocery store sells 5 cans of soup for $2.50.

   a. What is the rate in cans per dollar?

   b. What is the rate in dollars per can?

   c. What is the price of 10 cans of soup?

   d. Miss Baker has $15.00. What is the greatest number of cans of soup she can buy?

5. Eddie runs 2 laps around the gym every 3 minutes. Ryan says that Eddie runs at a rate of 1.5 laps per minute. Do you agree or disagree with Ryan? Use a ratio table or double number line to justify your reasoning.

6. Noah earns $30.00 for dog sitting for 4 days. Noah determines his rate in dollars per day and thinks the unit rate is 30. Do you agree or disagree with Noah? Justify your reasoning.

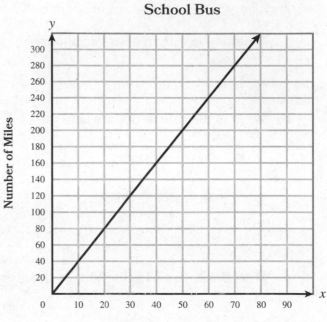

School Bus

Number of Miles (y-axis)

Number of Gallons of Fuel (x-axis)

7. The graph shows the ratio relationship between the number of miles a school bus travels and the number of gallons of fuel the school bus uses. Which statements appear to be true? Choose all that apply.

A. The bus travels 4 miles per gallon of fuel.

B. The bus uses 4 gallons of fuel per mile.

C. The bus uses 40 gallons of fuel for every 160 miles.

D. The bus travels $\frac{1}{4}$ mile per gallon of fuel.

E. The bus uses $\frac{1}{4}$ gallon of fuel per mile.

Remember

For problems 8 and 9, divide. Write the quotient and the remainder on separate lines.

8. $6,245 \div 7$

   Quotient: _____

   Remainder: _____

9. $9,371 \div 4$

   Quotient: _____

   Remainder: _____

10. Sasha and Julie make applesauce. Sasha uses 15 apples for every 3 pints of applesauce she makes. Julie uses 24 apples for every 6 pints of applesauce she makes. Who uses more apples per pint of applesauce? How do you know?

11. Round 152.096 to the indicated place value.

    a. hundredth

    b. tenth

    c. one

    d. ten

Name

Date

Comparing Rates

Drumming to the Beat

Which Is Faster?

The Perfect Tempo

Name _____   Date _____

1. The tables represent Kayla's and Toby's text messaging rates. Who sends more text messages per week? Explain.

Kayla

Number of Weeks	Number of Text Messages Sent
1	750
3	2,250

Toby

Number of Weeks	Number of Text Messages Sent
2	1,250
4	2,500

2. The tables show the number of ounces of blueberries used to make different numbers of muffins at Happy Muffin Shop and Sunny Muffin Shop. Lisa likes muffins with a lot of blueberries. Which muffin shop would you recommend to Lisa? Explain.

Happy Muffin Shop

Number of Ounces of Blueberries	Number of Muffins
12	24
24	48

Sunny Muffin Shop

Number of Ounces of Blueberries	Number of Muffins
60	80
90	120

3. Bus A travels 10 miles in 15 minutes. Bus B travels 8 miles in 10 minutes. Both buses start traveling at the same time. At these rates, which bus travels 50 miles first? Create two tables to show how you know.

4. Yuna earns $266.00 grooming 14 poodles. Tyler earns $180.00 grooming 9 poodles. Who earns less per poodle? Explain your reasoning.

5. Car wash A washes 5 cars in 1 hour. Car wash B washes 1 car in 15 minutes. Which company washes cars at a faster rate? Complete the double number lines to support your answer.

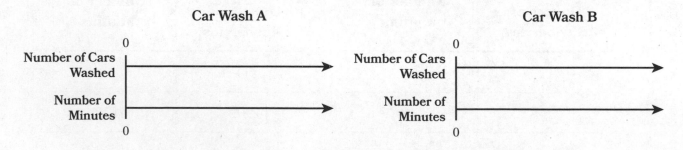

6. Sana makes 4 sandwiches in 8 minutes. Kelly says Sana's rate is 0.5 sandwiches per minute. Ryan says Sana's rate is 2 sandwiches per minute. Who is correct? Create two tables to show how you know.

Remember

For problems 7 and 8, divide. Write the quotient and the remainder on separate lines.

7. $6{,}874 \div 3$

Quotient: _____

Remainder: _____

8. $10{,}580 \div 6$

Quotient: _____

Remainder: _____

9. A hot air balloon travels at a constant speed of 15 miles per hour.

   a. Interpret the meaning of 15 miles per hour in this situation.

   b. Complete the ratio table.

Number of Hours	Number of Miles
1	
2	
3	
4	

   c. Use the ratio table from part (b) to create a double number line that models this situation.

10. Which pairs of fractions and decimals are equivalent? Choose all that apply.

   A. $\frac{2}{10}$ and 0.2

   B. $\frac{7}{10}$ and 0.07

   C. $\frac{3}{100}$ and 0.3

   D. $\frac{6}{100}$ and 0.06

   E. $\frac{19}{100}$ and 0.19

Name   Date

Using Rates to Convert Units

_12_ eggs per dozen

_100_ pennies per dollar

_4_ quarts per gallon

_1000_ meters per kilometer

Unit Conversions as Rates

1. Select three rates from the table of unit conversions. State the unit rate for each.

Using Rates to Convert Across Measurement Systems

2. Miss Baker wants to put a border around her class bulletin board. She measures the bulletin board in inches and finds that it measures 60 inches by 72 inches.

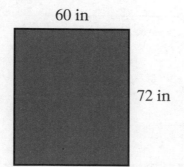

60 in

72 in

a. What is the perimeter of Miss Baker's bulletin board in inches?

b. Miss Baker arrives at the teacher supply store only to find that the border for sale is measured in centimeters. How many centimeters of border does Miss Baker need so she has enough to put around her bulletin board? Create a ratio table or another representation to support your answer.

c. Miss Baker also needs to buy sand for a science project. The project requires at least 5 pounds of sand. The store sells 40-ounce bags of sand. Is one 40-ounce bag of sand enough for the project? If not, how many 40-ounce bags of sand does Miss Baker need? Create a ratio table or another representation to support your answer.

3. A typical cruising speed for an airplane is 900 kilometers per hour. One kilometer is about 0.62 miles. How many miles per hour is the typical cruising speed for an airplane? Create a double number line or another representation to support your answer.

Using Rates to Solve Multi-Step Problems

4. Water flows from a garden hose at a rate of 91.2 liters per minute.

   a. A pool holds 8,640 gallons of water. How many minutes will it take to fill the pool with the garden hose? Create a ratio table or another representation to support your answer. Use 1 gallon ≈ 3.8 liters.

   b. How many hours will it take to fill the pool?

5. A restaurant cooks 10 pounds of roast beef to serve in sandwiches. Each sandwich has 100 grams of roast beef. Each pound is about 450 grams. How many sandwiches can the restaurant prepare with 10 pounds of roast beef?

Measurement Conversions Reference Table

Length	Weight	Capacity
1 inch = 2.54 centimeters	1 pound = 16 ounces	1 cup = 8 fluid ounces
1 meter ≈ 39.37 inches	1 pound ≈ 0.45 kilograms	1 pint = 2 cups
1 mile = 5,280 feet	1 kilogram ≈ 2.20 pounds	1 quart = 2 pints
1 mile = 1,760 yards	1 ton = 2,000 pounds	1 gallon = 4 quarts
1 mile ≈ 1.61 kilometers		1 gallon ≈ 3.79 liters
1 kilometer ≈ 0.62 miles		1 liter ≈ 0.26 gallons
		1 liter = 1,000 cubic centimeters

Name _____ Date _____

Water flows from a faucet at a rate of 1.7 gallons per minute.

a. At what rate does the water flow from the faucet in liters per minute?
    Use 1 gallon ≈ 3.79 liters. Round your answer to the nearest tenth.

b. At this rate, about how long would it take to fill a 500-liter fish tank? Round your answer to the
    nearest minute.

Name _____ Date _____

Use the grade 6 Measurement Conversions Reference Table.

For problems 1–11, fill in the blank to complete the unit conversion. If necessary, round to the nearest hundredth.

1. 7 ft = _____ in

2. 100 yd = _____ ft

3. 25 m = _____ cm

4. 4.34 km ≈ _____ mi

5. 96 oz = _____ lb

6. 2 mi ≈ _____ km

7. 3 in = _____ cm

8. 5 gal ≈ _____ L

9. 15 L = _____ gal

10. 6 g = _____ mg

11. 22 lb ≈ _____ kg

12. Miss Baker can walk 1 mile in 20 minutes. At that rate, how many miles can she walk per hour?

13. A typical midsize car weighs 1.5 tons. It has a gas tank that holds 15 gallons of gas.

    a. How many quarts does the gas tank hold?

    b. How many pounds does the car weigh?

14. Ryan buys a 2-kilogram bag of trail mix for a hike. He wants to make 5-ounce bags to share with his hiking friends. There are approximately 35 ounces in 1 kilogram. How many 5-ounce bags can Ryan make?

15. A great white shark can swim at a top speed of 40 kilometers per hour.

    a. How many meters per hour can the shark swim at its top speed?

    b. How many miles per hour can the shark swim at its top speed? Round your answer to the nearest whole number.

    c. Use the rate you found in part (b) to find the number of miles the shark can swim in 12 minutes at its top speed.

    d. How many miles per minute can the shark swim at its top speed?

Remember

For problems 16 and 17, divide.

16. $8,320 \div 40$

17. $11,700 \div 50$

18. Kayla texts 135 words in 3 minutes. What is the rate that she texts in words per minute? What is the unit rate?

19. Toby has 7 meters of fabric. He uses the fabric to make 10 same-size pillowcases. How much fabric in meters does Toby use for each pillowcase?

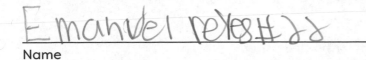

Name _____   Date _____

Solving Rate Problems

Family Road Trips

The Evans family, the Perez family, and the Chan family are taking road trips. Use the information given to answer at least one question at each station.

Station 1

Determine the best deal by finding the lowest price per unit of the products each family purchases.

a. The table shows the price each family pays for sunscreen.

Evans Family	Perez Family	Chan Family
6-ounce bottle $12.00	Two 4.2-ounce bottles $15.96	10-ounce bottle $19.50

[Handwritten work:]

Evans: $\frac{\$}{12} \frac{oz}{6} \div 6 = \frac{2}{1}$ 8/oz

Perez: $\frac{\$}{15.96} \frac{oz}{8.4} \div 8.4$ = 1.90/oz   1.90

Chan: $\frac{\$}{19.50} \frac{oz}{10} \div 10 = 1.95$ 1.95/1

b. The table shows the price each family pays for bottled water.

Evans Family	Perez Family	Chan Family
12-pack of 1-liter bottles $15.60	12-pack of 0.5-liter bottles $7.68	10-pack of 500-milliliter bottles $6.45

c. The table shows the price each family pays for trail mix.

Evans Family	Perez Family	Chan Family
6-pack of 150-gram bags $18.00	One 1-kilogram bag $15.00	Two 1-pound bags $22.50

Station 2

Each part provides the number of miles and the number of minutes for the beginning of each family's drive.

a. The Evans family drives 21 miles in the first 20 minutes of their trip. What is their speed in miles per hour during this time? Create a representation to show your work.

b. The Perez family drives 16.5 miles in the first 15 minutes of their trip. What is their speed in miles per hour during this time? Create a representation to show your work.

c. The Chan family drives 35 miles in the first 30 minutes of their trip. What is their speed in miles per hour during this time? Create a representation to show your work.

Station 3

Each part provides information about the gasoline each family buys. Use the given rate and the given quantity to determine the unknown quantity.

a. The Evans family stops at a gas station that charges $3.00 per gallon of gasoline. How many gallons of gasoline does the Evans family buy if they spend a total of $27.00?

b. The Perez family stops at a gas station and spends a total of $35.00 on 10 gallons of gasoline. How much does this gas station charge per gallon of gasoline?

c. The Chan family stops at a gas station that charges $3.25 per gallon of gasoline. What is the total amount in dollars the Chan family spends if they buy 8 gallons of gasoline?

Station 4

Each part requires you to convert from one unit of measurement to another. Use the information to answer the following questions.

a. The Evans children walk to the end of a fishing pier and back. The fishing pier is 1,320 feet in length. How many kilometers do the Evans children walk? Round to the nearest hundredth if necessary.

b. One of the Perez children sees a sand crab run 2 meters in 16 seconds. What rate in centimeters per second does the sand crab run?

c. One of the Chan children sees a hermit crab crawl 12 inches in 10 seconds. What rate in meters per minute does the hermit crab crawl? Round to the nearest hundredth if necessary.

Station 5

Each part provides rates in other situations on the road trips. Use the information to answer the following questions.

a. Mrs. Evans buys 6 small beach towels at a rate of $5.00 per towel. She buys 3 large beach towels at a different rate. She spends a total of $54.00 on beach towels. What rate in dollars per towel does Mrs. Evans pay for the large beach towels?

b. Mr. Perez stops to buy 9 gallons of gasoline at a rate of $3.00 per gallon. He stops a second time and buys 10 gallons of gasoline at a different rate. He spends a total of $58.00 on gasoline. What rate in dollars per gallon does Mr. Perez pay for gasoline at the second stop?

c. For the first 3 nights of the Chan family's stay, the hotel charges a rate of $125.00 per night. The hotel charges a different rate for the last 2 nights. The hotel charges a total of $585.00 for this 5-night stay. What rate in dollars per night does the hotel charge for the last 2 nights of the Chan family's stay?

Name _____

Date _____

Scott runs at a rate of 5 minutes per kilometer. His goal is to run a 50-kilometer race in under 4 hours. Will he meet his goal if he runs at his current rate? How do you know?

Name _____   Date _____

1. Lisa types at a rate of 40 words per minute.

    a. At this rate, how many words can Lisa type per hour?

    b. At this rate, how many words can Lisa type per second?

    c. If Lisa types at this same rate, how many minutes will it take her to type 320 words?

2. Scott is buying kale chips for his lunches. He can buy a pack of six $\frac{1}{2}$-ounce bags of kale chips for $9.00, or he can buy one 5-ounce bag of kale chips for $12.50. Which option costs less per ounce of kale chips? Explain.

3. A high-speed train travels at a rate of 90 miles per hour. At this rate, how many miles does the high-speed train travel in $4\frac{1}{2}$ hours?

4. The tank in a lawnmower holds 3 quarts of gasoline. If gasoline costs $3.00 per gallon, what is the total cost in dollars to fill the lawnmower's empty tank?

5. Two trampoline parks charge different rates. Trampoline park A charges $4.00 for every $\frac{1}{2}$ hour of jump time. Trampoline park B charges $20.00 for 3 hours of jump time. If you plan to visit for 3 hours, which trampoline park charges the better rate? Explain.

6. Store A sells a $\frac{1}{2}$-gallon size of juice for $2.65. This store sells the same juice in a 1-gallon size for $5.50.

   a. Which size of juice at store A, the $\frac{1}{2}$-gallon size or the 1-gallon size, has the better price per gallon? Explain.

   b. Store B sells the same juice in a 1-liter bottle for $1.50. Does the juice from store B have a better price per gallon than the juice from store A? Explain. Use 1 gallon ≈ 3.79 liters.

7. The running times and distances of four runners are listed. Put the runners in order from the slowest speed to the fastest speed in minutes per mile. Use 1 kilometer ≈ 0.62 miles.

   • Blake runs 2 miles in 17 minutes.

   • Kayla runs 3 miles in 27 minutes.

   • Noah runs 2.5 miles in 20 minutes.

   • Ryan runs 5 kilometers in 31 minutes.

8. Jada tutors science for 4 hours at a rate of $15.00 per hour. She tutors math for 2 hours at a different rate. She earns a total of $100.00 for these hours to tutor science and math. What rate in dollars per hour does Jada charge to tutor math?

9. At a bookfair, Riley buys 4 novels at a rate of $6.00 per book. She also buys 3 comic books at a different rate. Riley spends a total of $40.50 at the bookfair. What rate in dollars per book does she pay for the comic books?

Remember

For problems 10 and 11, divide.

10. $12{,}240 \div 20$

11. $16{,}020 \div 30$

12. Lacy and Tara bike to a campground 30 miles away. Lacy bikes 3 miles in 15 minutes. Tara bikes 5 miles in 20 minutes. Lacy and Tara leave at the same time and from the same location. At these rates, who reaches the campground first? Show how you know.

13. Tyler has 5 same-size granola bars to share equally with Leo and Sasha. If each person gets the same amount, how many granola bars do Tyler, Leo, and Sasha each get?

Name _____   Date _____

Solving Multi-Step Rate Problems

A Horse, of Course

1. A horse runs a 7,920-foot race in 150 seconds.

   a. What is the horse's rate in feet per second?

   b. What is the horse's rate in feet per minute?

   c. What is the horse's rate in feet per hour?

   d. What is the horse's rate in miles per hour?

Sending a Horse to the Moon

2. A horse runs 36 miles per hour. The distance from Earth to the moon is about 238,855 miles. How many days would it take a horse to run from Earth to the moon? Assume the horse runs at a constant speed. Assume there is actually a road from Earth to the moon.

Your Turn

3. Choose a method of travel from the list. The distance from Earth to the moon is about 238,855 miles. How many weeks would it take to travel from Earth to the moon? Assume the chosen traveler or mode of transportation moves at the given constant speed. As needed, assume there is actually a road from Earth to the moon.

A person running $\frac{1}{10}$ mile per minute	A scooter going 792 feet per minute
A sloth crawling 4 meters per minute	A train going 95,000 meters per hour
A peregrine falcon diving 88 meters per second	A dirt bike going 3,800 centimeters per second
A cheetah running 120 kilometers per hour	A fire truck going 55 feet per second
A Galápagos tortoise walking 984 feet per hour	Roller skates going 4,224 inches per minute

Name   Date

Modeling in *A Story of Ratios*

Read

Read the problem all the way through. Ask yourself:

- What is this problem asking me to find?

Then reread a chunk at a time. As you reread, ask yourself:

- What do I know?

Model the situation, possibly with tables, graphs, diagrams, and equations.

Represent

Represent the problem by using your chosen model. Ask yourself:

- What labels do I use on the table, graph, or diagram?
- How should I define the variables?

As you work, ask yourself:

- Are the known and the unknown clear in the model?

Add to or revise your model as necessary.

Solve

Solve the problem. Determine whether your result appears to be a correct solution. Ask yourself:

- Does my answer make sense?
- Does my result answer the question?

If not, revise your model or create a new one. Then ask yourself these questions again using your new result.

Summarize

Summarize your result and be ready to justify your reasoning.

Name

Date

A horsefly flies at a rate of 90 miles per hour. The distance from Earth to the moon is about 238,855 miles. Kayla says, "At that rate, it would take a horsefly less than 16 weeks to fly to the moon."

Do you agree or disagree with Kayla? Justify your answer. Round to the nearest tenth if necessary.

Name _____ Date _____

Solving Multi-Step Rate Problems

In this lesson, we

- converted both quantities of a rate to different units of measurement.
- solved multi-step rate problems.

Example

A professional runner runs a 100-yard race in 9.58 seconds. For parts (a)–(f), round answers to the hundredths place if necessary.

a. What is the runner's rate in yards per second?

$$100 \div 9.58 \approx 10.44$$

The runner's rate is about 10.44 yards per second.

Number of Seconds	Number of Yards
1	10.44
9.58	100

÷9.58   ÷9.58

b. What is the runner's rate in feet per second?

10.44 yards per second

1 yard = 3 feet

$$3 \times 10.44 = 31.32$$

The runner's rate is about 31.32 feet per second.

Number of Yards	Number of Feet
1	3
10.44	31.32

×10.44   ×10.44

c. What is the runner's rate in feet per minute?

31.32 feet per second

60 seconds = 1 minute

$$31.32 \times 60 = 1{,}879.2$$

The runner's rate is about 1,879.2 feet per minute.

Number of Minutes	Number of Feet
1	31.32
60	1,879.2

×60   ×60

d. What is the runner's rate in feet per hour?

$$1{,}879.2 \text{ feet per minute}$$

$$60 \text{ minutes} = 1 \text{ hour}$$

$$1{,}879.2 \times 60 = 112{,}752$$

The runner's rate is about 112,752 feet per hour.

Number of Minutes	Number of Feet
1	1,879.2
60	112,752

×60   ×60

e. What is the runner's rate in miles per hour?

$$112{,}752 \text{ feet per hour}$$

$$5{,}280 \text{ feet} = 1 \text{ mile}$$

$$112{,}752 \div 5{,}280 \approx 21.35$$

The runner's rate is about 21.35 miles per hour.

Number of Feet	Number of Miles
5,280	1
112,752 ÷5,280	21.35

÷5,280

f. If the runner could maintain this rate, how long would it take him to run a 3-mile race?

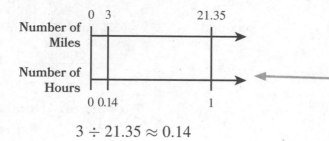

$$3 \div 21.35 \approx 0.14$$

It would take the runner about 0.14 hours to run a 3-mile race.

Number of Miles	Number of Hours
21.35	1
3 ÷21.35	0.14

÷21.35

Sometimes, it helps to convert units so that the answer makes more sense. Because there are 60 minutes in 1 hour, you can multiply 60 and 1 each by 0.14 to find that there are 8.4 minutes in 0.14 hours. This means it would take the runner about 8.4 minutes to run a 3-mile race at this rate.

Name _____ Date _____

1. Rain falls at a rate of $\frac{4}{5}$ inches per hour during a storm. At this rate, how many inches of rain fall in 10 hours?

2. A school bus leaves a middle school at 2:00 p.m.

   a. The bus takes the basketball team to a game 30 miles away. If the bus travels at a rate of 40 miles per hour, what time does the team arrive at the game?

   b. On the way back, the bus travels at a rate of 35 miles per hour. How far does the bus travel in 15 minutes?

3. Leo runs the 40-yard dash in 4.84 seconds. Round answers to the nearest hundredth as necessary.

   a. What is Leo's rate in yards per second?

   b. What is Leo's rate in feet per second?

   c. What is Leo's rate in feet per minute?

   d. What is Leo's rate in feet per hour?

   e. What is Leo's rate in miles per hour?

f. Imagine that Leo could actually maintain this pace for a long time. How many hours would it take him to run the 277-mile length of the Grand Canyon?

4. Adesh competes in a triathlon. The race has three parts: a swim, a bike ride, and a run.

a. Adesh swims at a rate of 50 meters per minute. He completes the swim part of the triathlon in 30 minutes. How far does he swim in kilometers?

b. Adesh bikes at a rate of 32 kilometers per hour. How many minutes does it take him to complete the 40-kilometer bike ride?

c. There are two 10-minute transitions during the triathlon. One is between the swim and the bike ride. The other is between the bike ride and the run. Including the two 10-minute transitions, Adesh finishes the triathlon in 3 hours and 5 minutes. Find Adesh's rate in kilometers per hour for the 10-kilometer run part of the triathlon.

Remember

For problems 5 and 6, divide.

5. $25,060 \div 70$

6. $75,000 \div 60$

7. An antelope can run as fast as 60 miles per hour.

a. At that rate, how many miles can an antelope run in 15 minutes?

b. At that rate, how many feet can an antelope run in 2 minutes?

8. When traveling to Mexico, Jada exchanged her 10 US dollars for 200 Mexican pesos. What is the exchange rate between US dollars and Mexican pesos? Choose all that apply.

A. 20 dollars per peso

B. 20 pesos per dollar

C. $\frac{1}{20}$ peso per dollar

D. $\frac{1}{20}$ dollar per peso

A Generous Donation

When it comes to money matters—such as taxes, charitable donations, and more—we often think in terms of percents.

Consider $1. If you have $1,000, then a single dollar wouldn't feel like much. But if you have only $2, then that single dollar would feel like a pretty big deal. It would be half of what you own!

When you're evaluating whether an amount of money is "a lot" or "a little," it's not enough to consider the money itself. You also need a point of comparison. That's why percents are so common—because, as ratios in disguise, they help us make comparisons.

Name _____ Date _____

Introduction to Percents

Charging Batteries

What Is the Charge?

Greater Than 100%

Name

Date

1. The shaded squares in the grid show the occupied seats in a movie theater.

a. What is the ratio of the number of occupied seats to the total number of seats?

b. What fraction of the seats are occupied?

c. What percent of the seats are occupied?

d. What percent of the seats are not occupied?

2. Consider 9%.

   a. Shade the grid to represent 9%.

   b. Write 9% as a fraction.

   c. Write 9% as a decimal.

3. Consider $\frac{2}{5}$.

   a. Write an equivalent fraction with a denominator of 100.

   b. Write $\frac{2}{5}$ as a decimal.

   c. Write $\frac{2}{5}$ as a percent.

4. Complete the table.

Grid	Percent	Decimal	Fraction
	17%		
		0.7	
			$\frac{7}{100}$

5. Consider the grids. One whole is a single 10×10 grid.

a. Represent the shaded area as a fraction.

b. Represent the shaded area as a decimal.

c. Represent the shaded area as a percent.

6. Yuna says that 40% of the diagram is not shaded. Lacy says that 4% of the diagram is not shaded. Who is correct? Explain how you know.

7. Leo says that 11% of the diagram is shaded. Adesh says that 55% of the diagram is shaded. Who is correct? Explain how you know.

Remember

For problems 8 and 9, divide. Write the quotient and the remainder on separate lines.

8. $6,052 \div 20$

9. $5,410 \div 30$

10. Sasha types at a rate of 35 words per minute.

    a. At this rate, how many words does Sasha type in one half-hour?

    b. At this rate, how many words does Sasha type per second?

For problems 11 and 12, list all the factors for the number shown.

11. 12

12. 40

Name _____ Date _____

Finding the Percent

About 2,000 years ago, the Roman Empire was a vast territory. To maintain the empire, taxes were collected from its citizens. Initially, the tax system was very unfair because it relied on people called tax farmers to collect taxes in an area called a province. These tax farmers prepaid the estimated taxes for their province and then kept any extra money they collected from the citizens. Tax farmers were able to get rich quickly this way. Roman emperor Augustus Caesar stopped the process of tax farming and taxed Roman citizens more fairly.

Roman citizens earned income and paid taxes with silver coins called *denarii*.

1. Imagine that you are the emperor of Rome. You want to tax the citizens fairly. How much would you tax each of the following citizens? Explain your thinking.

   a. Citizen A earns 225 denarii per year.

   b. Citizen B earns 375 denarii per year.

   c. Citizen C earns 3,750 denarii per year.

2. The symbol ‰ is called *permille*. The symbol ‰₀ is called *permyriad*. How can 80% be expressed in permille? How can it be expressed in permyriad?

Using Tape Diagrams to Model Percents

3. The tape diagram represents the remaining charge of a cell phone's battery.

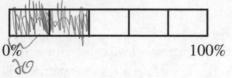

0%   100%

a. What percent of the battery's charge does each unit of the tape diagram represent? Explain.

b. What fraction represents the battery's remaining charge? What fraction represents the battery's charge that has been used?

c. What percent of the battery's charge remains? What percent of the battery's charge has been used?

4. The following students estimate the number of hours their cell phone battery will last when it has a whole charge.

whole 160%

a. Ryan's phone battery lasts 10 hours. How many hours will Ryan's phone last when the battery's remaining charge is 40%? Draw a tape diagram to determine your answer.

10 hours

40   100%

Ryan phone has 4 hour left.

b. Kayla's phone battery lasts 15 hours. How many hours will Kayla's phone last when the battery's remaining charge is 40%? Draw a tape diagram to determine your answer.

c. Riley's phone battery lasts 8 hours. How many hours will Riley's phone last when the battery's remaining charge is 40%? Draw a tape diagram to determine your answer.

d. Lacy's phone battery lasts 6 hours. How many hours will Lacy's phone last when the battery's remaining charge is 40%? Draw a tape diagram to determine your answer.

5. Complete the table with the numbers from parts (a)–(d) in problem 4.

Remaining Amount of Battery's Charge (hours)	Total Amount of Battery's Charge (hours)

For problems 6–9, solve by using any method.

6. 13 out of 20 is what percent?

7. 18 out of 36 is what percent?

8. What percent is 24 out of 80?

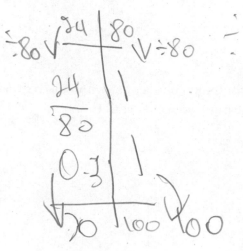

9. What percent is 17 out of 85?

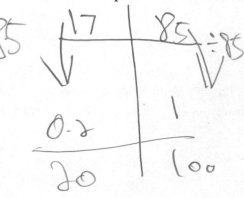

Using Double Number Lines to Model Percents

10. Noah sets a goal to eat at least 80 grams of protein each day. Today, he eats 92 grams. What percent of his goal does Noah eat today? Draw a double number line to show your thinking.

For problems 11–14, solve by using any method.

11. 18 is what percent of 15?

12. 450 out of 150 is what percent?

13. 21 out of 20 is what percent?

14. What percent is 675 out of 500?

Name _____   Date _____

Blake receives a gift of $80.00. He spends $60.00 on a new game and $8.00 on snacks.

a. Draw a double number line to represent the percent of Blake's gift money he spends.

b. What percent of Blake's gift money does he spend on a new game?

c. What percent of Blake's gift money does he spend on snacks?

d. What percent of Blake's gift money does he have remaining?

Name _____ Date _____

1. Riley has $200.00. She spends $60.00 on a video game.

   a. What percent of her money does Riley spend? Draw a diagram to show your thinking.

   b. What percent of her money does Riley have remaining?

For problems 2–7, determine the unknown percent by using any method.

2. 15 out of 25 is what percent?

3. What percent is 9 out of 10?

4. What percent of 40 is 12?

5. 45 is _____ percent of 60.

6. What percent of 30 is 45?

7. 250 is what percent of 125?

For problems 8–13, solve by using any method.

8. There are 25 students in Kayla's math class. There are 32 students in her physical education class.

   a. 32 is what percent of 25?

   b. What does the answer to part (a) represent in this situation?

   c. 12 students from Kayla's math class say that math is their favorite subject. What percent of the students in Kayla's math class say that math is their favorite subject?

   d. 24 students in Kayla's physical education class say that soccer is their favorite game. What percent of the students in Kayla's physical education class say that soccer is their favorite game?

9. Toby drives 170 miles of his 500-mile road trip.

    a. What percent of his road trip does Toby drive?

    b. What percent of his road trip does Toby have left to drive?

10. Animal clinic A has 4 cats out of a total of 10 animals. Animal clinic B has 12 cats out of a total of 25 animals. Which clinic has the greater percent of cats?

11. In her hometown, Yuna pays $0.80 in tax on a $10.00 purchase. When visiting a beach town, she pays $0.50 in tax on a $5.00 purchase. Which town charges the greater tax percent on purchases?

12. Noah estimates that his maximum heart rate is 208 beats per minute. Noah's teacher tells him that his target heart rate for physical activity is between 70% and 85% of his maximum heart rate. After running one mile, Noah's heart rate is 156 beats per minute.

When Noah ran one mile, did he meet his target heart rate? Explain.

13. Scott earns $12.00 per hour at his after-school job. After a pay raise, Scott makes $15.00 per hour.

a. 15 is what percent of 12?

b. What does your answer to part (a) represent in this situation?

Remember

For problems 14 and 15, divide. Write the quotient and the remainder on separate lines.

14. $6,244 \div 40$

15. $9,560 \div 50$

16. On Saturday, Leo earns a total of $70.00 for mowing lawns and doing chores. He mows 3 lawns and earns $15.00 per lawn. He does chores for 2 hours. What rate in dollars per hour does Leo earn for doing chores?

17. 25 is a multiple of which numbers? Choose all that apply.

    A. 1

    B. 5

    C. 10

    D. 25

    E. 50

LESSON 24

Finding a Part

1. A group of students want their school to start an art club. Jada, Tyler, Blake, and Lisa each gather data from a survey to determine the number of students at the school who support the creation of an art club. Use the double number lines provided to determine the number of students in each set of data who support the creation of an art club.

Jada's Data

75% of 76 students support the creation of an art club.

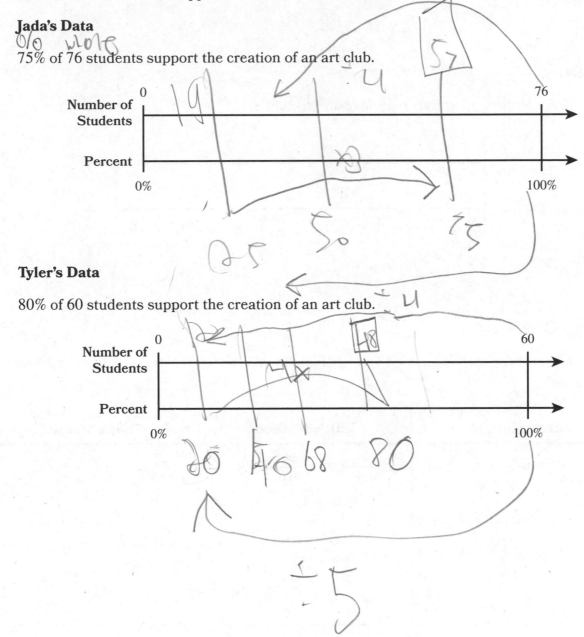

Tyler's Data

80% of 60 students support the creation of an art club.

Blake's Data

15% of 360 students support the creation of an art club.

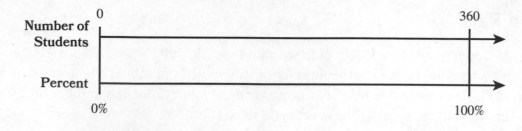

Lisa's Data

70% of 90 students support the creation of an art club.

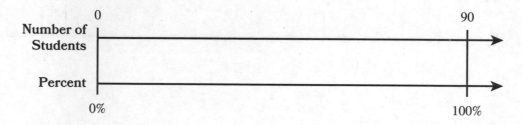

More Than One Way

2. Lisa, Julie, and Toby each calculated 16% of 40 but in different ways. Use their work to answer parts (a)–(g).

Lisa's Method	Julie's Method	Toby's Method
10% of 40 is 4. 5% of 40 is 2. 1% of 40 is 0.4. $10\% + 5\% + 1\% = 16\%$ $4 + 2 + 0.4 = 6.4$ 16% of 40 is 6.4	$\frac{16}{100} \times 40 = \frac{640}{100} = 6.4$ 16% of 40 is 6.4.	1% of 40 is 0.4. $16 \times 0.4 = 6.4$ So 16% of 40 is 6.4.

a. Explain how Lisa calculated 16% of 40.

b. Use Lisa's method to calculate 31% of 50.

c. What is another percent of 40 you could calculate by using Lisa's method? Calculate that percent.

d. Explain how Julie calculated 16% of 40.

e. Use Julie's method to calculate 3% of 80.

f. How are Lisa's and Toby's methods similar? How are they different?

g. Use Toby's method to calculate 8% of 5.

3. Noah noticed that he can calculate 10% of 40 by dividing 40 by 10. To calculate 32% of 40, he divides 40 by 32. Does his method work? Explain.

4. The girls' soccer team at a middle school has 25 players.

    a. Five of the players are seventh-grade students. Find the percent of seventh graders on the team. Explain or show your thinking.

    b. Of the 25 girls on the team, 60% started playing soccer in elementary school. How many of the girls on the team started playing soccer in elementary school? Explain or show your thinking.

    c. The coach says that 15% of the players will miss an upcoming game. Why must this statement be false?

Would You Rather?

For problems 5–10, answer the question and justify your choice mathematically.

5. Would you rather have 10% of $6.00 or 80% of 90 cents?

6. Would you rather help 15% of 80 people or 75% of 20 people?

7. Would you rather eat 20% of 240 spicy peppers or 2% of 1,000 spicy peppers?

8. Would you rather eat 11% of 300 jelly beans or 90% of 40 jelly beans?

9. Would you rather clean 9% of 2,000 dishes or 14% of 1,500 dishes?

10. Would you rather drink 24% of a 20-ounce smoothie or 41% of a 15-ounce smoothie?

Double Number Lines

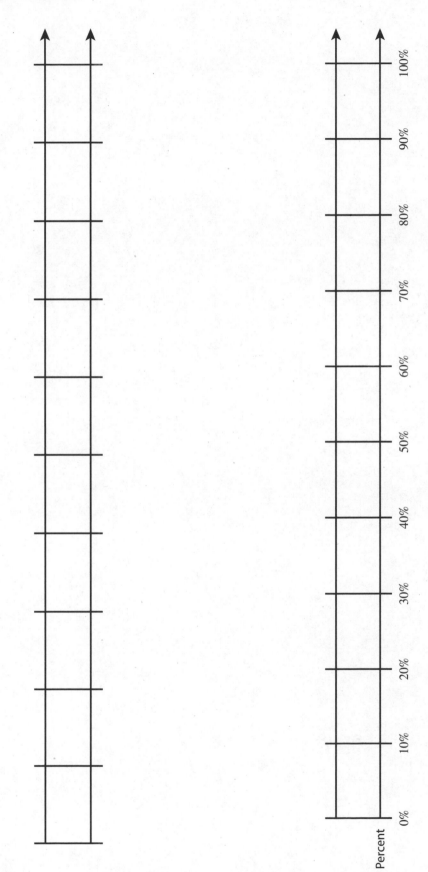

Name _____   Date _____

A giant jellyfish weighs 440 pounds. If 95% of the jellyfish's weight is water, how many pounds of the jellyfish's weight is water? Justify your answer.

Name _____ Date _____

1. How many pounds is 50% of 16 pounds?

2. How many miles is 25% of 44 miles?

3. How many miles is 100% of 44 miles?

4. How many kilograms is 10% of 96 kilograms?

5. How many meters is 20% of 24 meters?

6. How many minutes is 5% of 60 minutes?

7. How many dollars is 1% of $560.00?

8. Which of the following statements are true? Choose all that apply.

   A. 50% of 346 is equal to 0.5×346.

   B. 25% of 210 is equal to $210 \div 4$.

   C. 20% of 460 is equal to $460 \div 5$.

   D. 10% of 642 is equal to $\frac{1}{10} \times 642$.

   E. 1% of 198 is equal to $\frac{1}{1,000} \times 198$.

9. There are 60 people in a room. If 65% of the people in the room have brown hair, how many people have brown hair? Use the double number line to support your answer.

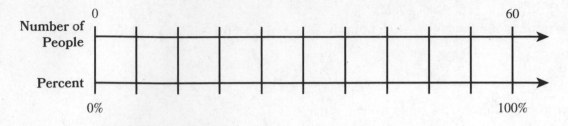

10. There are 45 marbles in a jar, and 80% of them are blue. How many blue marbles are in the jar?

11. Of the 32 students in a class, 75% participate in after-school activities. How many students in the class participate in after-school activities?

12. There are 300 people at a museum, and 18% of them are adults. How many adults are at the museum?

13. Ryan receives a gift of $75.00. He donates 24% of this money to his favorite charity. How much money does he donate?

14. Which amount is greater, 15% of 20 or 20% of 15? Explain.

Remember

For problems 15 and 16, divide. Write the quotient and the remainder on separate lines.

15. $4,725 \div 30$

16. $1,565 \div 60$

17. Consider 62%.

a. Shade the grid to represent 62%.

b. Write 62% as a fraction.

c. Write 62% as a decimal.

18. Which numbers are multiples of 6? Choose all that apply.

A. 6

B. 16

C. 18

D. 36

E. 40

F. 72

$$6 = 25$$

$$25 = \frac{1}{4} = 6$$

Name _____   Date _____

Finding the Whole

1. There are 4 colors of balloons. Sana and Tara each have a different total number of balloons. What is the least total number of balloons that each girl could have?

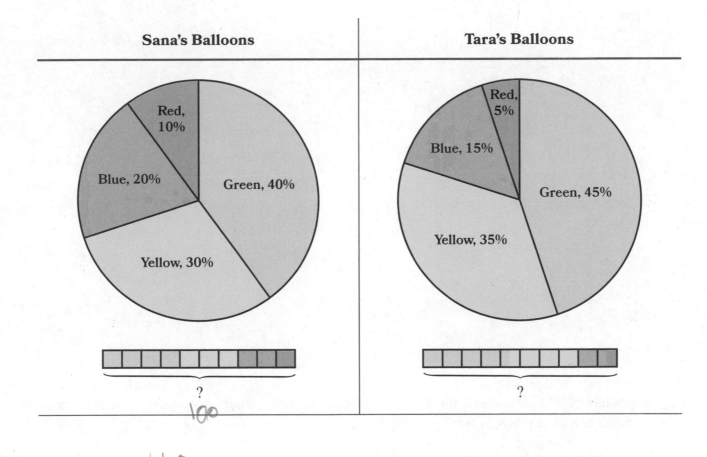

Sana's Balloons

Red, 10%
Blue, 20%
Green, 40%
Yellow, 30%

?

100

Tara's Balloons

Red, 5%
Blue, 15%
Green, 45%
Yellow, 35%

?

40

What Is the Whole?

2. If 6 balloons is 30% of the total number of balloons, what is the total number of balloons? Draw a tape diagram or a double number line to support your answer.

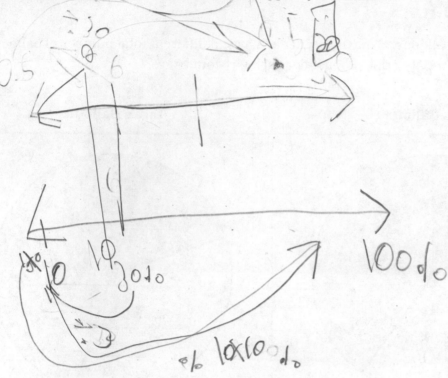

3. Kayla has 12 red balloons. If this is 3% of the total number of balloons she has, what is the total number of balloons Kayla has?

Solving Multi-Step Percent Problems

4. Toby has raised $225.00 for his favorite charity. This is 60% of his goal. How much more money does Toby need to raise to meet his goal?

5. Lisa has raised $750.00 for her favorite charity. This is 120% of her goal. How much extra money has Lisa raised so far?

6. The sixth, seventh, and eighth grades each raised money for their middle school. The sixth grade raised 40% of the total amount raised. The seventh grade raised 22% of the total amount raised. The eighth grade raised the rest. If the eighth grade raised $1,368.00, what is the total amount of money raised by the three grades?

7. Yuna and Scott were the only two candidates who ran for sixth-grade class president. Yuna received 45% of the votes. Scott received 18 more votes than Yuna. How many votes did Scott receive?

Name _____   Date _____

A team has raised $300.00 for new uniforms, which is 60% of the total amount of money they need to raise. What is the total amount of money the team needs to raise? Justify your answer.

7. Lisa has saved $33.00. This amount is 55% of her savings goal. How much money does Lisa have left to save to reach her savings goal?

8. Sasha has saved $40.00. This amount is 125% of her savings goal. How much extra money has Sasha saved?

9. A survey at a middle school reports that 48% of students prefer to read mysteries, 32% of students prefer to read science fiction, and the rest of the students prefer to read nonfiction. If 25 students prefer to read nonfiction, how many students prefer to read mysteries?

10. At a company, 20% of the employees are part-time employees. There are 33 more full-time employees than part-time employees. What is the total number of employees at this company?

Remember

For problems 11 and 12, divide. Write the quotient and the remainder on separate lines.

11. $4,831 \div 30$

12. $9,510 \div 50$

13. Noah has $200.00. He spends $120.00 on a new coat.

    a. What percent of his money does Noah spend on a new coat?

    b. What percent of his money does Noah have left?

14. Use the double number line to determine which of the following statements are true. Choose all that apply.

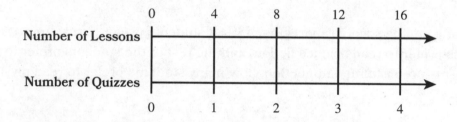

    A. The ratio of the number of lessons to the number of quizzes is 4 : 1.

    B. For every 1 quiz, there are 4 lessons.

    C. The ratio of the number of quizzes to the number of lessons is 12 : 3.

    D. For every 8 lessons, there are 2 quizzes.

    E. For every 12 lessons, there are 4 quizzes.

Name _____ Date _____

Solving Percent Problems

Mysterious Donations

1. Mrs. A has $200.00. Mrs. B has $400.00. Mrs. C has $600.00. One of these women wants to donate 1% of her money to you. Another woman wants to donate 2% of her money to you. The last woman wants to donate 3% of her money to you. However, you do not know who wants to donate which percent.

a. What are all the possible total amounts of money you could receive? Show or explain why these are all possible total amounts of money.

b. What is the greatest total amount of money you could receive?

c. What is the least total amount of money you could receive?

Cafeteria Calculations

2. The school cafeteria manager asks your math class to help design a lunch menu. You can choose from the foods in the list provided. The manager gives you the following requirements:

*   The lunch menu should have a total number of calories that is no fewer than 600 and no more than 700.

*   The lunch menu should have one main dish, one or two sides, and one drink.

*   Between 45% and 65% of the total number of calories should come from carbohydrates.

*   Between 20% and 35% of the total number of calories should come from fat.

*   Between 10% and 35% of the total number of calories should come from protein.

There are 4 calories per gram of carbohydrates, 9 calories per gram of fat, and 4 calories per gram of protein.

Nutrition Tables

Main Dishes

Food Item	Number of Calories	Amount of Carbohydrates (grams)	Amount of Fat (grams)	Amount of Protein (grams)
Bean burrito	277	37	9	12
Spaghetti and meatballs	414	50	14	22
Chicken nuggets	227	16	11	16
Turkey hot dog	241	24	13	7
Beef nachos	412	40.5	20	17.5
Cheese pizza	351	36	15	18
Turkey and cheese sandwich	288	27.5	12	17.5
Cheeseburger	343	28	17	19.5

Side Dishes

Food Item	Number of Calories	Amount of Carbohydrates (grams)	Amount of Fat (grams)	Amount of Protein (grams)
Apple	59	14	0.2	0.3
Carrot sticks	28	6	0	1
String cheese	59	1	3	7
Corn	85	17	1	2
Graham crackers	95	16	3	1
Brown rice	145	31	1	3
Cherry tomatoes	20	4	0	1
Mashed potatoes and gravy	207	19	7	17

Drinks

Drink Item	Number of Calories	Amount of Carbohydrates (grams)	Amount of Fat (grams)	Amount of Protein (grams)
Chocolate milk, nonfat	140	26	0	9
Milk, 1%	131	16	3	10
Water	0	0	0	0

Name _____   Date _____

1. Blake uses 60% of the amount of his paycheck to pay bills. After paying all of his bills, Blake has $320.00 of his paycheck remaining. What is the amount of Blake's paycheck? Justify your answer.

2. Kayla has $8,836.00 in her savings account. The bank gives Kayla 5% of the amount of money in the account as a customer bonus. What amount of money does the bank give Kayla? Justify your answer.

Name _____ Date _____

Solving Percent Problems

In this lesson, we

- used our understanding of percents to solve multi-step percent problems.

Examples

1. Yuna has run 24 miles this week, which is 60% of her weekly running goal. What is the total number of miles Yuna needs to run to meet her weekly running goal? Use a double number line to show your thinking.

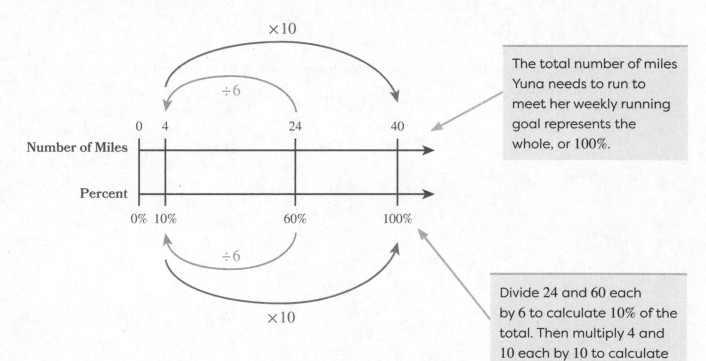

The total number of miles Yuna needs to run to meet her weekly running goal represents the whole, or 100%.

Divide 24 and 60 each by 6 to calculate 10% of the total. Then multiply 4 and 10 each by 10 to calculate 100% of the total.

Yuna needs to run 40 miles to meet her weekly running goal.

2. The chess club has a goal to raise $500.00 for charity. So far, the club has raised $380.00. What percent of the goal has the chess club raised so far?

$$\frac{380}{500} = \frac{38}{50} = \frac{76}{100}$$

The chess club has raised 76% of its goal so far.

One method to find an unknown percent is to write the part out of the whole as a fraction. Then find an equivalent fraction with a denominator of 100.

Another method is to divide the part, 380, by the whole, 500, to get 0.76. This is the decimal form of 76%.

3. A basketball team wins 70% of its games. If the team plays a total of 20 games, how many games does the team win?

$$\frac{70}{100} = 0.7$$

$$0.7 \times 20 = 14$$

The answer to this question is the same as the solution to the problem, What is 70% of 20?

The team wins 14 games.

Name _____ Date _____

For problems 1–5, solve by using any method.

1. Kayla runs laps for a fitness challenge in physical education class. After Kayla runs 18 laps, her friends shout, "Yay! You're 45% done!" How many laps is Kayla trying to run?

2. The soccer team raises $600.00 during a fundraiser. The team's goal is to raise a total of $750.00. What percent of the team's goal does the team raise during the fundraiser?

3. The theater club set a goal to sell 48 theater tickets in one day. At the end of that day, the theater teacher says that the club sold 175% of its goal amount. How many tickets did the club sell?

4. A store offers both a 40% off coupon and a $10.00 off coupon. Only one coupon may be used for each item purchased.

   a. Which coupon gives the lower price of the item? Mark the better coupon for each item in the table.

Item Purchased	40% Off Coupon	$10.00 Off Coupon
$140.00 Bike		
$15.00 Scarf		
$75.00 Winter Coat		
$12.00 Book		

   b. Show how you made your choices in part (a). Write the final price of each item after the better coupon has been used.

   Bike:

   Scarf:

Winter coat:

Book:

c. Lisa chooses a shirt to buy from this store. No matter which coupon she uses, the price of the shirt will be discounted by the same amount. What is the price of the shirt? Explain how you know.

5. Toby considers buying three different pairs of running shoes: a $20.00 pair, a $40.00 pair, and a $60.00 pair. He has three coupons: one for 10% off one item, one for 20% off one item, and one for 25% off one item. Assume that Toby buys at least one pair of running shoes and uses a coupon for each pair of shoes he buys.

What is the least amount of money Toby could pay? What is the greatest amount? Explain.

Remember

For problems 6 and 7, divide. Write the quotient and the remainder on separate lines.

6. $5,264 \div 52$

7. $8,630 \div 65$

8. Some bats can fly at a rate of 85 miles per hour. Yuna says, "At that rate, it would take a bat less than 2 weeks to fly the 29,401-mile distance around Earth at the equator." Do you agree or disagree with Yuna? Justify your answer. Round to the nearest tenth if necessary.

9. Which statements accurately describe the ratio relationship shown in the diagram? Choose all that apply.

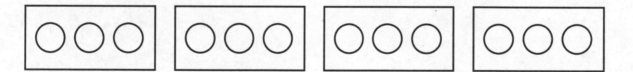

A. For every 4 boxes, there are 12 circles.

B. For every 4 boxes, there are 3 circles.

C. For every 4 circles, there are 3 boxes.

D. For every 3 circles, there is 1 box.

E. For every 1 box, there are 3 circles.

Mixed Practice 1

Name _____   Date _____

For problems 1 and 2, list all the factors of the given number.

1. 12

2. 40

3. Evaluate $55 \div 5 \times (40 - 4 + 2)$.

For problems 4–7, write the value that makes the measurements equivalent.

4. 7 meters = _____ centimeters

5. 4,000 milliliters = _____ liters

6. 9 kilometers = _____ meters

7. 18 kilograms = _____ grams

For problems 8 and 9, use the figure to complete the question. The measure of ∠*JKM* is 170°.

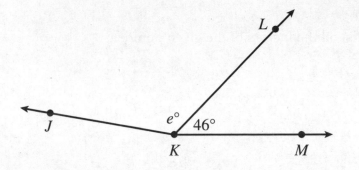

8. Write an equation using *e* to find the measure of ∠*JKL*.

9. What is the measure in degrees of ∠*JKL*?

10. For one week, students in Mr. Sharma's class collect water in rain gauges at home. At the end of the week, each student reports the height of the water in their gauge to the rest of the class. The data are shown in the table.

Student	Height of Water (inches)
Jada	$2\frac{3}{4}$
Kelly	$\frac{5}{8}$
Toby	$\frac{6}{8}$
Adesh	$1\frac{3}{4}$
Eddie	$2\frac{1}{4}$
Tara	$1\frac{3}{4}$
Sasha	$2\frac{1}{4}$
Leo	$1\frac{1}{4}$
Scott	$2\frac{2}{8}$
Riley	$1\frac{3}{8}$

a. Make a line plot to display the data.

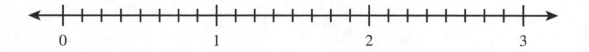

b. How many students report a water height of more than $1\frac{1}{2}$ inches?

c. What is the difference in inches between the greatest water height and the least water height?

11. Lacy builds a rectangular prism with the same volume as the prism shown.

What could be the measurements of Lacy's rectangular prism?

A. The length is 4 units, the width is 3 units, and the height is 3 units.

B. The length is 5 units, the width is 1 unit, and the height is 6 units.

C. The length is 5 units, the width is 2 units, and the height is 2 units.

D. The length is 3 units, the width is 5 units, and the height is 3 units.

12. Add.

$$5\frac{3}{8} + 1\frac{6}{10}$$

13. Match each division expression to its equivalent multiplication expression

$2 \div \frac{1}{5}$ 3×4

$3 \div \frac{1}{4}$ 4×3

$4 \div \frac{1}{3}$ 2×5

Mixed Practice 2

Name _____ Date _____

1. Find the quotient and the remainder.

$$429 \div 8$$

2. Which of the following values are equivalent to 2.306? Choose all that apply.

   A. Two and three hundred six thousandths

   B. Two and thirty-six thousandths

   C. $2\frac{306}{1,000}$

   D. $(2 \times 1) + \left(3 \times \frac{1}{10}\right) + \left(6 \times \frac{1}{1,000}\right)$

   E. $(2 \times 1) + \left(3 \times \frac{1}{10}\right) + \left(6 \times \frac{1}{100}\right)$

   F. $\frac{2,306}{1,000}$

3. Multiply.

$$712 \times 308$$

4. Kelly buys a T-shirt for $7.99 and a pair of jeans for $15.50. He gives the clerk $30.00. How much change does Kelly receive?

5. How many lines of symmetry does the given shape have?

   A. 1

   B. 2

   C. 3

   D. 4

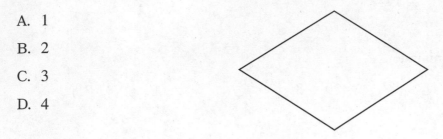

6. Plot the points in the coordinate plane. Connect the points in the order that they are given.

   (3, 5), (5, 2), (6, 4), (7, 2), (9, 5)

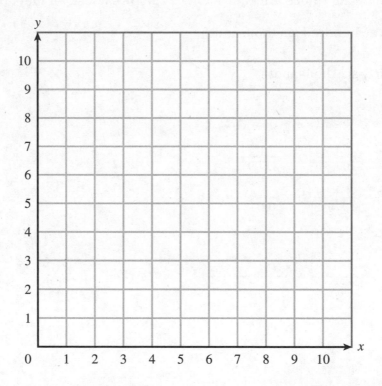

7. Use the word bank to complete true statements about quadrilaterals. Do not use any words more than once.

square	rhombus	parallelogram	rectangle

   a. A rectangle and a square are both examples of a _____.

   b. A _____ is always a rectangle.

   c. A _____ is not always a square.

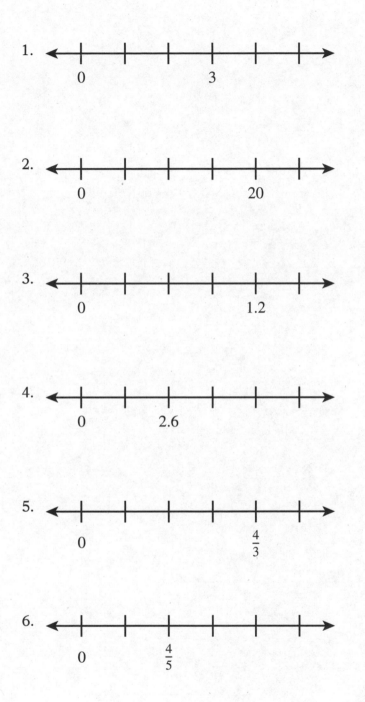

1.
    0   3

2.
    0   20

3.
    0   1.2

4.
    0   2.6

5.
    0                $\frac{4}{3}$

6.
    0      $\frac{4}{5}$

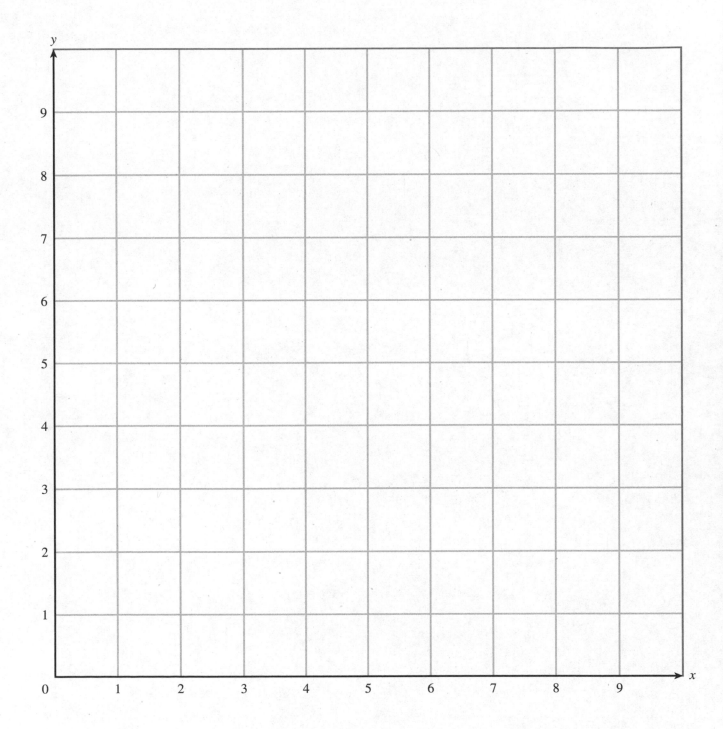

Sprint

Compare the fractions by using $<$, $>$, or $=$.

1.	$\frac{1}{3}$ ◯ $\frac{1}{9}$
2.	$\frac{1}{4}$ ◯ $\frac{3}{12}$

Compare the fractions by using <, >, or =.

Number Correct: _____

1.	$\frac{1}{2}\ \bigcirc\ \frac{1}{8}$
2.	$\frac{1}{2}\ \bigcirc\ \frac{1}{6}$
3.	$\frac{1}{2}\ \bigcirc\ \frac{1}{4}$
4.	$\frac{1}{6}\ \bigcirc\ \frac{1}{4}$
5.	$\frac{1}{8}\ \bigcirc\ \frac{1}{4}$
6.	$\frac{1}{8}\ \bigcirc\ \frac{1}{6}$
7.	$\frac{1}{2}\ \bigcirc\ \frac{3}{10}$
8.	$\frac{1}{2}\ \bigcirc\ \frac{4}{10}$
9.	$\frac{1}{2}\ \bigcirc\ \frac{5}{10}$
10.	$\frac{1}{2}\ \bigcirc\ \frac{7}{10}$
11.	$\frac{1}{2}\ \bigcirc\ \frac{9}{10}$
12.	$\frac{1}{3}\ \bigcirc\ \frac{8}{9}$
13.	$\frac{1}{3}\ \bigcirc\ \frac{4}{9}$
14.	$\frac{1}{3}\ \bigcirc\ \frac{3}{9}$
15.	$\frac{1}{3}\ \bigcirc\ \frac{2}{9}$
16.	$\frac{1}{3}\ \bigcirc\ \frac{3}{12}$
17.	$\frac{1}{3}\ \bigcirc\ \frac{4}{12}$
18.	$\frac{1}{3}\ \bigcirc\ \frac{6}{12}$

19.	$\frac{2}{9}\ \bigcirc\ \frac{2}{3}$
20.	$\frac{2}{9}\ \bigcirc\ \frac{2}{5}$
21.	$\frac{2}{9}\ \bigcirc\ \frac{2}{7}$
22.	$\frac{2}{9}\ \bigcirc\ \frac{2}{10}$
23.	$\frac{8}{9}\ \bigcirc\ \frac{8}{13}$
24.	$\frac{8}{9}\ \bigcirc\ \frac{8}{11}$
25.	$\frac{8}{9}\ \bigcirc\ \frac{10}{11}$
26.	$\frac{8}{9}\ \bigcirc\ \frac{5}{6}$
27.	$\frac{6}{9}\ \bigcirc\ \frac{5}{6}$
28.	$\frac{6}{9}\ \bigcirc\ \frac{4}{6}$
29.	$\frac{3}{9}\ \bigcirc\ \frac{2}{6}$
30.	$\frac{4}{9}\ \bigcirc\ \frac{1}{3}$
31.	$\frac{5}{12}\ \bigcirc\ \frac{1}{3}$
32.	$\frac{5}{12}\ \bigcirc\ \frac{2}{3}$
33.	$\frac{7}{12}\ \bigcirc\ \frac{2}{3}$
34.	$\frac{7}{18}\ \bigcirc\ \frac{2}{3}$
35.	$\frac{7}{18}\ \bigcirc\ \frac{5}{6}$
36.	$\frac{10}{18}\ \bigcirc\ \frac{5}{12}$

B

Number Correct: _____

Improvement: _____

Compare the fractions by using <, >, or =.

1.	$\frac{1}{10}$ ◯ $\frac{1}{2}$
2.	$\frac{1}{8}$ ◯ $\frac{1}{2}$
3.	$\frac{1}{6}$ ◯ $\frac{1}{2}$
4.	$\frac{1}{4}$ ◯ $\frac{1}{2}$
5.	$\frac{1}{4}$ ◯ $\frac{1}{8}$
6.	$\frac{1}{8}$ ◯ $\frac{1}{6}$
7.	$\frac{2}{10}$ ◯ $\frac{1}{2}$
8.	$\frac{3}{10}$ ◯ $\frac{1}{2}$
9.	$\frac{5}{10}$ ◯ $\frac{1}{2}$
10.	$\frac{6}{10}$ ◯ $\frac{1}{2}$
11.	$\frac{9}{10}$ ◯ $\frac{1}{2}$
12.	$\frac{3}{9}$ ◯ $\frac{1}{3}$
13.	$\frac{5}{9}$ ◯ $\frac{1}{3}$
14.	$\frac{6}{9}$ ◯ $\frac{1}{3}$
15.	$\frac{8}{9}$ ◯ $\frac{1}{3}$
16.	$\frac{6}{12}$ ◯ $\frac{1}{3}$
17.	$\frac{4}{12}$ ◯ $\frac{1}{3}$
18.	$\frac{3}{12}$ ◯ $\frac{1}{3}$

19.	$\frac{2}{7}$ ◯ $\frac{2}{3}$
20.	$\frac{2}{7}$ ◯ $\frac{2}{5}$
21.	$\frac{2}{7}$ ◯ $\frac{2}{9}$
22.	$\frac{2}{7}$ ◯ $\frac{2}{12}$
23.	$\frac{7}{8}$ ◯ $\frac{7}{12}$
24.	$\frac{7}{8}$ ◯ $\frac{7}{9}$
25.	$\frac{7}{8}$ ◯ $\frac{3}{4}$
26.	$\frac{7}{8}$ ◯ $\frac{10}{11}$
27.	$\frac{5}{6}$ ◯ $\frac{7}{8}$
28.	$\frac{5}{6}$ ◯ $\frac{10}{12}$
29.	$\frac{4}{9}$ ◯ $\frac{2}{3}$
30.	$\frac{4}{9}$ ◯ $\frac{1}{3}$
31.	$\frac{5}{12}$ ◯ $\frac{1}{4}$
32.	$\frac{5}{12}$ ◯ $\frac{3}{4}$
33.	$\frac{5}{12}$ ◯ $\frac{5}{6}$
34.	$\frac{7}{24}$ ◯ $\frac{1}{3}$
35.	$\frac{6}{24}$ ◯ $\frac{1}{3}$
36.	$\frac{6}{24}$ ◯ $\frac{1}{4}$

Sprint

Convert each measurement to the given unit.

1.	2 yd = _____ ft	
2.	$4\frac{1}{4}$ ft = _____ in	

A

Number Correct: _____

Convert each measurement to the given unit.

1.	1 yd = _____ ft	
2.	2 yd = _____ ft	
3.	3 yd = _____ ft	
4.	10 yd = _____ ft	
5.	5 yd = _____ ft	
6.	1 ft = _____ in	
7.	2 ft = _____ in	
8.	3 ft = _____ in	
9.	5 ft = _____ in	
10.	10 ft = _____ in	
11.	$\frac{1}{2}$ ft = _____ in	
12.	12 in = _____ ft	
13.	24 in = _____ ft	
14.	48 in = _____ ft	
15.	6 in = _____ ft	
16.	36 in = _____ ft	
17.	36 in = _____ yd	
18.	72 in = _____ yd	
19.	18 in = _____ yd	
20.	1 yd = _____ in	
21.	2 yd = _____ in	
22.	$\frac{1}{2}$ yd = _____ in	

23.	$1\frac{1}{3}$ yd = _____ ft	
24.	$1\frac{2}{3}$ yd = _____ ft	
25.	$2\frac{2}{3}$ yd = _____ ft	
26.	$3\frac{1}{3}$ yd = _____ ft	
27.	$6\frac{2}{3}$ yd = _____ ft	
28.	$1\frac{1}{2}$ ft = _____ in	
29.	$3\frac{1}{2}$ ft = _____ in	
30.	$4\frac{1}{4}$ ft = _____ in	
31.	$4\frac{3}{4}$ ft = _____ in	
32.	$5\frac{1}{3}$ ft = _____ in	
33.	36 in = _____ ft	
34.	42 in = _____ ft	
35.	48 in = _____ ft	
36.	66 in = _____ ft	
37.	36 in = _____ yd	
38.	18 in = _____ yd	
39.	48 in = _____ yd	
40.	90 in = _____ yd	
41.	120 in = _____ yd	
42.	120 in = _____ ft	
43.	$4\frac{1}{2}$ yd = _____ ft	
44.	$4\frac{1}{2}$ yd = _____ in	

B

Number Correct: _____

Improvement: _____

Convert each measurement to the given unit.

1.	1 ft = _____ in	
2.	2 ft = _____ in	
3.	4 ft = _____ in	
4.	8 ft = _____ in	
5.	10 ft = _____ in	
6.	$\frac{1}{2}$ ft = _____ in	
7.	1 yd = _____ ft	
8.	3 yd = _____ ft	
9.	6 yd = _____ ft	
10.	12 yd = _____ ft	
11.	10 yd = _____ ft	
12.	36 in = _____ yd	
13.	18 in = _____ yd	
14.	12 in = _____ yd	
15.	1 yd = _____ in	
16.	2 yd = _____ in	
17.	$\frac{1}{2}$ yd = _____ in	
18.	12 in = _____ ft	
19.	36 in = _____ ft	
20.	48 in = _____ ft	
21.	60 in = _____ ft	
22.	6 in = _____ ft	

23.	$\frac{1}{3}$ ft = _____ in	
24.	$1\frac{1}{2}$ ft = _____ in	
25.	$1\frac{1}{3}$ ft = _____ in	
26.	$3\frac{1}{3}$ ft = _____ in	
27.	$4\frac{1}{2}$ ft = _____ in	
28.	$5\frac{1}{2}$ ft = _____ in	
29.	$2\frac{1}{3}$ yd = _____ ft	
30.	$2\frac{2}{3}$ yd = _____ ft	
31.	$3\frac{2}{3}$ yd = _____ ft	
32.	$5\frac{1}{3}$ yd = _____ ft	
33.	$6\frac{2}{3}$ yd = _____ ft	
34.	48 in = _____ yd	
35.	54 in = _____ yd	
36.	90 in = _____ yd	
37.	24 in = _____ ft	
38.	40 in = _____ ft	
39.	42 in = _____ ft	
40.	54 in = _____ ft	
41.	96 in = _____ ft	
42.	96 in = _____ yd	
43.	$5\frac{1}{2}$ yd = _____ ft	
44.	$5\frac{1}{2}$ yd = _____ in	

Sprint

Write each fraction as a decimal.

1.	$\dfrac{2}{10}$	
2.	$\dfrac{7}{100}$	

Number Correct: _____

Write each fraction as a decimal.

1.	1 tenth	
2.	2 tenths	
3.	$\frac{3}{10}$	
4.	$\frac{4}{10}$	
5.	$\frac{5}{10}$	
6.	$\frac{6}{10}$	
7.	$\frac{7}{10}$	
8.	$\frac{8}{10}$	
9.	$\frac{9}{10}$	
10.	$\frac{10}{10}$	
11.	$\frac{11}{10}$	
12.	$\frac{12}{10}$	
13.	$\frac{15}{10}$	
14.	$\frac{17}{10}$	
15.	$\frac{21}{10}$	
16.	$\frac{24}{10}$	
17.	$\frac{26}{10}$	
18.	$\frac{32}{10}$	

19.	1 hundredth	
20.	4 hundredths	
21.	$\frac{6}{100}$	
22.	$\frac{9}{100}$	
23.	$\frac{10}{100}$	
24.	$\frac{12}{100}$	
25.	$\frac{17}{100}$	
26.	$\frac{21}{100}$	
27.	$\frac{24}{100}$	
28.	$\frac{36}{100}$	
29.	$\frac{48}{100}$	
30.	$\frac{67}{100}$	
31.	$\frac{134}{100}$	
32.	$\frac{221}{100}$	
33.	$\frac{274}{100}$	
34.	$\frac{308}{100}$	
35.	$\frac{430}{100}$	
36.	$\frac{701}{100}$	

B

Number Correct: _____

Improvement: _____

Write each fraction as a decimal.

1.	1 tenth	
2.	2 tenths	
3.	$\frac{1}{10}$	
4.	$\frac{2}{10}$	
5.	$\frac{4}{10}$	
6.	$\frac{5}{10}$	
7.	$\frac{7}{10}$	
8.	$\frac{8}{10}$	
9.	$\frac{9}{10}$	
10.	$\frac{10}{10}$	
11.	$\frac{12}{10}$	
12.	$\frac{14}{10}$	
13.	$\frac{15}{10}$	
14.	$\frac{18}{10}$	
15.	$\frac{23}{10}$	
16.	$\frac{24}{10}$	
17.	$\frac{29}{10}$	
18.	$\frac{35}{10}$	

19.	1 hundredth	
20.	3 hundredths	
21.	$\frac{1}{100}$	
22.	$\frac{8}{100}$	
23.	$\frac{10}{100}$	
24.	$\frac{15}{100}$	
25.	$\frac{20}{100}$	
26.	$\frac{22}{100}$	
27.	$\frac{28}{100}$	
28.	$\frac{39}{100}$	
29.	$\frac{43}{100}$	
30.	$\frac{68}{100}$	
31.	$\frac{125}{100}$	
32.	$\frac{230}{100}$	
33.	$\frac{294}{100}$	
34.	$\frac{311}{100}$	
35.	$\frac{447}{100}$	
36.	$\frac{721}{100}$	

Sprint

Find the unknown numerator.

1.	$\dfrac{6}{20} = \dfrac{}{10}$	
2.	$\dfrac{4}{5} = \dfrac{}{100}$	

A

Find the unknown numerator.

1.	$\frac{1}{5} = \frac{}{10}$	
2.	$\frac{2}{5} = \frac{}{10}$	
3.	$\frac{3}{5} = \frac{}{10}$	
4.	$\frac{4}{5} = \frac{}{10}$	
5.	$\frac{1}{2} = \frac{}{10}$	
6.	$\frac{10}{20} = \frac{}{10}$	
7.	$\frac{12}{20} = \frac{}{10}$	
8.	$\frac{14}{20} = \frac{}{10}$	
9.	$\frac{16}{20} = \frac{}{10}$	
10.	$\frac{18}{20} = \frac{}{10}$	
11.	$\frac{8}{20} = \frac{}{10}$	
12.	$\frac{6}{20} = \frac{}{10}$	
13.	$\frac{6}{30} = \frac{}{10}$	
14.	$\frac{5}{50} = \frac{}{10}$	
15.	$\frac{10}{50} = \frac{}{10}$	
16.	$\frac{20}{50} = \frac{}{10}$	
17.	$\frac{30}{50} = \frac{}{10}$	
18.	$\frac{40}{50} = \frac{}{10}$	

19.	$\frac{1}{10} = \frac{}{100}$	
20.	$\frac{3}{10} = \frac{}{100}$	
21.	$\frac{6}{10} = \frac{}{100}$	
22.	$\frac{8}{10} = \frac{}{100}$	
23.	$\frac{5}{10} = \frac{}{100}$	
24.	$\frac{1}{2} = \frac{}{100}$	
25.	$\frac{1}{4} = \frac{}{100}$	
26.	$\frac{3}{4} = \frac{}{100}$	
27.	$\frac{2}{5} = \frac{}{100}$	
28.	$\frac{3}{5} = \frac{}{100}$	
29.	$\frac{12}{20} = \frac{}{100}$	
30.	$\frac{15}{20} = \frac{}{100}$	
31.	$\frac{18}{20} = \frac{}{100}$	
32.	$\frac{36}{40} = \frac{}{100}$	
33.	$\frac{12}{40} = \frac{}{100}$	
34.	$\frac{24}{40} = \frac{}{100}$	
35.	$\frac{22}{40} = \frac{}{100}$	
36.	$\frac{32}{40} = \frac{}{100}$	

B

Number Correct: _____

Improvement: _____

Find the unknown numerator.

1.	$\dfrac{1}{2} = \dfrac{}{10}$	
2.	$\dfrac{10}{20} = \dfrac{}{10}$	
3.	$\dfrac{2}{20} = \dfrac{}{10}$	
4.	$\dfrac{6}{20} = \dfrac{}{10}$	
5.	$\dfrac{8}{20} = \dfrac{}{10}$	
6.	$\dfrac{12}{20} = \dfrac{}{10}$	
7.	$\dfrac{16}{20} = \dfrac{}{10}$	
8.	$\dfrac{20}{20} = \dfrac{}{10}$	
9.	$\dfrac{6}{30} = \dfrac{}{10}$	
10.	$\dfrac{12}{30} = \dfrac{}{10}$	
11.	$\dfrac{18}{30} = \dfrac{}{10}$	
12.	$\dfrac{24}{30} = \dfrac{}{10}$	
13.	$\dfrac{30}{30} = \dfrac{}{10}$	
14.	$\dfrac{4}{40} = \dfrac{}{10}$	
15.	$\dfrac{8}{40} = \dfrac{}{10}$	
16.	$\dfrac{16}{40} = \dfrac{}{10}$	
17.	$\dfrac{24}{40} = \dfrac{}{10}$	
18.	$\dfrac{40}{40} = \dfrac{}{10}$	

19.	$\dfrac{1}{10} = \dfrac{}{100}$	
20.	$\dfrac{2}{10} = \dfrac{}{100}$	
21.	$\dfrac{4}{10} = \dfrac{}{100}$	
22.	$\dfrac{6}{10} = \dfrac{}{100}$	
23.	$\dfrac{5}{10} = \dfrac{}{100}$	
24.	$\dfrac{1}{2} = \dfrac{}{100}$	
25.	$\dfrac{1}{4} = \dfrac{}{100}$	
26.	$\dfrac{2}{4} = \dfrac{}{100}$	
27.	$\dfrac{1}{5} = \dfrac{}{100}$	
28.	$\dfrac{2}{5} = \dfrac{}{100}$	
29.	$\dfrac{8}{20} = \dfrac{}{100}$	
30.	$\dfrac{15}{20} = \dfrac{}{100}$	
31.	$\dfrac{14}{20} = \dfrac{}{100}$	
32.	$\dfrac{28}{40} = \dfrac{}{100}$	
33.	$\dfrac{2}{40} = \dfrac{}{100}$	
34.	$\dfrac{12}{40} = \dfrac{}{100}$	
35.	$\dfrac{24}{40} = \dfrac{}{100}$	
36.	$\dfrac{30}{40} = \dfrac{}{100}$	

Sprint

Find the unknown factor.

1.	_____ × 10 = 100	
2.	5 × 5 × _____ = 100	

A

Number Correct: _____

Find the unknown factor.

1.	$10 \times \underline{\hspace{1cm}} = 100$	
2.	$\underline{\hspace{1cm}} \times 10 = 100$	
3.	$4 \times \underline{\hspace{1cm}} = 100$	
4.	$25 \times \underline{\hspace{1cm}} = 100$	
5.	$2 \times \underline{\hspace{1cm}} = 100$	
6.	$50 \times \underline{\hspace{1cm}} = 100$	
7.	$5 \times \underline{\hspace{1cm}} = 100$	
8.	$20 \times \underline{\hspace{1cm}} = 100$	
9.	$\underline{\hspace{1cm}} \times 20 = 100$	
10.	$2 \times \underline{\hspace{1cm}} \times 10 = 100$	
11.	$5 \times 2 \times \underline{\hspace{1cm}} = 100$	
12.	$2 \times 5 \times \underline{\hspace{1cm}} = 100$	
13.	$\underline{\hspace{1cm}} \times 2 \times 5 = 100$	
14.	$\underline{\hspace{1cm}} \times 5 \times 2 = 100$	
15.	$10 \times 5 \times \underline{\hspace{1cm}} = 100$	
16.	$5 \times 10 \times \underline{\hspace{1cm}} = 100$	
17.	$5 \times \underline{\hspace{1cm}} \times 10 = 100$	
18.	$10 \times \underline{\hspace{1cm}} \times 5 = 100$	

19.	$2 \times 10 \times \underline{\hspace{1cm}} = 100$	
20.	$4 \times 5 \times \underline{\hspace{1cm}} = 100$	
21.	$4 \times \underline{\hspace{1cm}} \times 5 = 100$	
22.	$\underline{\hspace{1cm}} \times 2 \times 25 = 100$	
23.	$5 \times 5 \times \underline{\hspace{1cm}} = 100$	
24.	$\underline{\hspace{1cm}} \times 5 \times 5 = 100$	
25.	$25 \times 2 \times \underline{\hspace{1cm}} = 100$	
26.	$2 \times \underline{\hspace{1cm}} \times 25 = 100$	
27.	$10 \times 10 = 2 \times \underline{\hspace{1cm}} \times 10$	
28.	$10 \times \underline{\hspace{1cm}} \times 2 = 20 \times 5$	
29.	$5 \times 5 \times 2 \times 2 = 4 \times \underline{\hspace{1cm}}$	
30.	$4 \times 5 \times 5 = 2 \times \underline{\hspace{1cm}} \times 5$	
31.	$25 \times 4 = \underline{\hspace{1cm}} \times 2 \times 5 \times 5$	
32.	$5 \times 2 \times 2 \times \underline{\hspace{1cm}} = 10 \times 10$	
33.	$25 \times 2 \times 2 = 5 \times \underline{\hspace{1cm}} \times 2$	
34.	$10 \times 2 \times 5 = \underline{\hspace{1cm}} \times 50$	
35.	$\underline{\hspace{1cm}} \times 2 \times 2 \times 5 = 100$	
36.	$4 \times 5 \times 5 = 10 \times \underline{\hspace{1cm}} \times 5$	

B

Number Correct: _____

Improvement: _____

Find the unknown factor.

1.	$2 \times$ _____ $= 100$	
2.	$50 \times$ _____ $= 100$	
3.	$4 \times$ _____ $= 100$	
4.	$25 \times$ _____ $= 100$	
5.	$10 \times$ _____ $= 100$	
6.	_____ $\times 10 = 100$	
7.	$2 \times$ _____ $\times 10 = 100$	
8.	$5 \times$ _____ $\times 10 = 100$	
9.	_____ $\times 2 \times 5 = 100$	
10.	_____ $\times 5 \times 2 = 100$	
11.	$5 \times 2 \times$ _____ $= 100$	
12.	$2 \times 5 \times$ _____ $= 100$	
13.	$10 \times 5 \times$ _____ $= 100$	
14.	$5 \times 10 \times$ _____ $= 100$	
15.	$5 \times$ _____ $= 100$	
16.	$20 \times$ _____ $= 100$	
17.	_____ $\times 20 = 100$	
18.	$10 \times$ _____ $\times 5 = 100$	

19.	$5 \times 5 \times$ _____ $= 100$	
20.	_____ $\times 5 \times 5 = 100$	
21.	$25 \times 2 \times$ _____ $= 100$	
22.	_____ $\times 2 \times 25 = 100$	
23.	$2 \times 10 \times$ _____ $= 100$	
24.	$4 \times 5 \times$ _____ $= 100$	
25.	$4 \times$ _____ $\times 5 = 100$	
26.	$2 \times$ _____ $\times 25 = 100$	
27.	$5 \times 20 = 10 \times$ _____ $\times 5$	
28.	$10 \times 2 \times$ _____ $= 4 \times 25$	
29.	$5 \times 4 \times 5 = 5 \times 2 \times$ _____	
30.	$4 \times 25 = 2 \times$ _____ $\times 5 \times 5$	
31.	$25 \times 4 = 5 \times 5 \times$ _____ $\times 2$	
32.	$2 \times 2 \times 5 \times$ _____ $= 10 \times 10$	
33.	$2 \times 25 \times 2 = 5 \times$ _____ $\times 2$	
34.	$10 \times 2 \times 5 = 50 \times$ _____	
35.	_____ $\times 5 \times 5 \times 2 = 100$	
36.	$5 \times 4 \times 5 = 10 \times$ _____	

Sprint

Convert each measurement to the given unit.

1.	$3\,\text{km} = \underline{\hspace{1cm}}\,\text{m}$	
2.	$7\,\text{m} = \underline{\hspace{1cm}}\,\text{cm}$	

A

Number Correct: _____

Convert each measurement to the given unit.

1.	1 km = _____ m		23.	6 km = _____ m	
2.	2 km = _____ m		24.	6.5 km = _____ m	
3.	3 km = _____ m		25.	6.35 km = _____ m	
4.	7 km = _____ m		26.	6.125 km = _____ m	
5.	5 km = _____ m		27.	9.054 km = _____ m	
6.	1 m = _____ cm		28.	7 m = _____ cm	
7.	2 m = _____ cm		29.	7.5 m = _____ cm	
8.	3 m = _____ cm		30.	7.48 m = _____ cm	
9.	9 m = _____ cm		31.	7.03 m = _____ cm	
10.	6 m = _____ cm		32.	7.035 m = _____ cm	
11.	12 m = _____ cm		33.	500 cm = _____ m	
12.	1,000 m = _____ km		34.	515 cm = _____ m	
13.	9,000 m = _____ km		35.	510 cm = _____ m	
14.	8,000 m = _____ km		36.	523.4 cm = _____ m	
15.	5,000 m = _____ km		37.	100 cm = _____ m	
16.	500 m = _____ km		38.	50 cm = _____ m	
17.	100 cm = _____ m		39.	75 cm = _____ m	
18.	200 cm = _____ m		40.	25 cm = _____ m	
19.	300 cm = _____ m		41.	125 cm = _____ m	
20.	900 cm = _____ m		42.	125 m = _____ km	
21.	1,000 cm = _____ m		43.	125 km = _____ m	
22.	10 cm = _____ m		44.	12.5 m = _____ cm	

B

Number Correct: _____

Improvement: _____

Convert each measurement to the given unit.

1.	1 km = _____ m	
2.	3 km = _____ m	
3.	5 km = _____ m	
4.	8 km = _____ m	
5.	9 km = _____ m	
6.	1 m = _____ cm	
7.	3 m = _____ cm	
8.	7 m = _____ cm	
9.	8 m = _____ cm	
10.	5 m = _____ cm	
11.	10 m = _____ cm	
12.	1,000 m = _____ km	
13.	6,000 m = _____ km	
14.	4,000 m = _____ km	
15.	2,000 m = _____ km	
16.	500 m = _____ km	
17.	100 cm = _____ m	
18.	300 cm = _____ m	
19.	500 cm = _____ m	
20.	800 cm = _____ m	
21.	1,200 cm = _____ m	
22.	10 cm = _____ m	

23.	4 km = _____ m	
24.	4.5 km = _____ m	
25.	4.25 km = _____ m	
26.	4.375 km = _____ m	
27.	8.154 km = _____ m	
28.	9 m = _____ cm	
29.	9.3 m = _____ cm	
30.	9.25 m = _____ cm	
31.	9.04 m = _____ cm	
32.	9.045 m = _____ cm	
33.	300 cm = _____ m	
34.	315 cm = _____ m	
35.	310 cm = _____ m	
36.	323.4 cm = _____ m	
37.	100 cm = _____ m	
38.	50 cm = _____ m	
39.	25 cm = _____ m	
40.	75 cm = _____ m	
41.	150 cm = _____ m	
42.	150 m = _____ km	
43.	150 km = _____ m	
44.	15 m = _____ cm	

Credits

Great Minds® has made every effort to obtain permission for the reprinting of all copyrighted material. If any owner of copyrighted material is not acknowledged herein, please contact Great Minds for proper acknowledgment in all future editions and reprints of this module.

Cover, Gustave Caillebotte (1848–1894). *Paris Street; Rainy Day*, 1877 Oil on canvas, 212.2 x 276.2 cm (83 1/2 x 108 3/4 in.). Charles H. and Mary F. S. Worcester Collection. (1964.336). The Art Institute of Chicago, Chicago, IL, U.S.A. Photo Credit: The Art Institute of Chicago/Art Resource, NY; page 66, Alena_D/Shutterstock.com; page 321, Eduardo Estellez/Shutterstock.com; All other images are the property of Great Minds.

For a complete list of credits, visit http://eurmath.link/media-credits.

Acknowledgments

Agnes P. Bannigan, Erik Brandon, Joseph T. Brennan, Beth Brown, Amanda H. Carter, Mary Christensen-Cooper, David Choukalas, Cheri DeBusk, Jill Diniz, Mary Drayer, Dane Ehlert, Scott Farrar, Kelli Ferko, Levi Fletcher, Krysta Gibbs, Winnie Gilbert, Julie Grove, Marvin E. Harrell, Stefanie Hassan, Robert Hollister, Rachel Hylton, Travis Jones, Raena King, Emily Koesters, Liz Krisher, Robin Kubasiak, Connie Laughlin, Alonso Llerena, Gabrielle Mathiesen, Maureen McNamara Jones, Bruce Myers, Marya Myers, Kati O'Neill, Ben Orlin, Darion Pack, Brian Petras, DesLey V. Plaisance, Lora Podgorny, Janae Pritchett, Bonnie Sanders, Deborah Schluben, Andrew Senkowski, Erika Silva, Ashley Spencer, Hester Sofranko, Danielle Stantoznik, Tara Stewart, Heidi Strate, James Tanton, Jessica Vialva, Carla Van Winkle, Caroline Yang

Trevor Barnes, Brianna Bemel, Adam Cardais, Christina Cooper, Natasha Curtis, Jessica Dahl, Brandon Dawley, Delsena Draper, Sandy Engelman, Tamara Estrada, Soudea Forbes, Jen Forbus, Reba Frederics, Liz Gabbard, Diana Ghazzawi, Lisa Giddens-White, Laurie Gonsoulin, Nathan Hall, Cassie Hart, Marcela Hernandez, Rachel Hirsh, Abbi Hoerst, Libby Howard, Amy Kanjuka, Ashley Kelley, Lisa King, Sarah Kopec, Drew Krepp, Crystal Love, Maya Márquez, Siena Mazero, Cindy Medici, Ivonne Mercado, Sandra Mercado, Brian Methe, Patricia Mickelberry, Mary-Lise Nazaire, Corinne Newbegin, Max Oosterbaan, Tamara Otto, Christine Palmtag, Andy Peterson, Lizette Porras, Karen Rollhauser, Neela Roy, Gina Schenck, Amy Schoon, Aaron Shields, Leigh Sterten, Mary Sudul, Lisa Sweeney, Samuel Weyand, Dave White, Charmaine Whitman, Nicole Williams, Glenda Wisenburn-Burke, Howard Yaffe

Talking Tool

Share Your Thinking	I know . . . . I did it this way because . . . . The answer is _____ because . . . . My drawing shows . . . .
Agree or Disagree	I agree because . . . . That is true because . . . . I disagree because . . . . That is not true because . . . . Do you agree or disagree with _____? Why?
Ask for Reasoning	Why did you . . . ? Can you explain . . . ? What can we do first? How is _____ related to _____?
Say It Again	I heard you say . . . . _____ said . . . . Another way to say that is . . . . What does that mean?

Thinking Tool

When I solve a problem or work on a task, I ask myself

Before

Have I done something like this before?

What strategy will I use?

Do I need any tools?

During

Is my strategy working?

Should I try something else?

Does this make sense?

After

What worked well?

What will I do differently next time?

At the end of each class, I ask myself

What did I learn?

What do I have a question about?